Disaster Management and Preparedness

Occupational Safety and Health Guide Series

Series Editor

Thomas D. Schneid
Eastern Kentucky University
Richmond, Kentucky

Published Titles

Creative Safety Solutions
by Thomas D. Schneid

Occupational Health Guide to Violence in the Workplace
by Thomas D. Schneid

Motor Carrier Safety: A Guide to Regulatory Compliance
by E. Scott Dunlap

Disaster Management and Preparedness
by Thomas D. Schneid and Larry R. Collins

Managing Workers' Compensation: A Guide to Injury Reduction and Effective Claim Management
by Keith R. Wertz and James J. Bryant

Physical Hazards of the Workplace
by Larry R. Collins and Thomas D. Schneid

Disaster Management and Preparedness

Thomas D. Schneid
Larry Collins

LEWIS PUBLISHERS
Boca Raton   London   New York   Washington, D.C.

Library of Congress Cataloging-in-Publication Data

Schneid, Thomas D.
  Disaster management and preparedness / Thomas D. Schneid and Larry Collins.
    p. cm.— (Occupational safety and health guide series)
  Includes bibliographical references and index.
  ISBN 1-56670-524-X
  1. Emergency management. 2. Emergency management—United States. I. Collins, Larry, 1952- II. Title. III. Series.
HV551.2 S39   2000
658.4'77—dc21   00-061252

This book contains information obtained from authentic and highly regarded sources. Reprinted material is quoted with permission, and sources are indicated. A wide variety of references are listed. Reasonable efforts have been made to publish reliable data and information, but the authors and the publisher cannot assume responsibility for the validity of all materials or for the consequences of their use.

Neither this book nor any part may be reproduced or transmitted in any form or by any means, electronic or mechanical, including photocopying, microfilming, and recording, or by any information storage or retrieval system, without prior permission in writing from the publisher.

The consent of CRC Press LLC does not extend to copying for general distribution, for promotion, for creating new works, or for resale. Specific permission must be obtained in writing from CRC Press LLC for such copying.

Direct all inquiries to CRC Press LLC, 2000 N.W. Corporate Blvd., Boca Raton, Florida 33431.

Trademark Notice: Product or corporate names may be trademarks or registered trademarks, and are used only for identification and explanation, without intent to infringe.

Visit the CRC Press Web site at www.crcpress.com

© 2000 by CRC Press LLC

No claim to original U.S. Government works
International Standard Book Number 1-56670-524-X
Library of Congress Card Number 00-061252
Printed in the United States of America   6 7 8 9 0
Printed on acid-free paper

About the Authors

Dr. Thomas D. Schneid is a tenured professor in the Department of Loss Prevention and Safety at Eastern Kentucky University and serves as the coordinator of the Fire and Safety Engineering program. He is also a founding member of the law firm of Schumann & Schneid, PLLC located in Richmond, Kentucky.

Dr. Schneid earned a B.S. in Education, M.S. and C.A.S. in Safety, M.S. in International Business, and Ph.D. in Environmental Engineering as well as his J.D. in law from West Virginia University and LL.M. (Graduate law) from the University of San Diego. He also earned an M.S. in International Business and a Ph.D. in Environmental Engineering. He is a member of the Bar for the U.S. Supreme Court, 6th Circuit Court of Appeals, and a number of federal districts as well as the Kentucky and West Virginia Bar Associations.

Dr. Schneid has authored and co-authored 15 texts on various safety, fire, EMS, and legal topics as well as over 100 articles. He was named one of the "Rising Stars in Safety" by *Occupational Hazards* magazine in 1997 and recently awarded the Program of Distinction Fellow by the Commonwealth of Kentucky and EKU.

Dr. Larry Collins (A.S., B.S., M.S., Ed.D.) joined the LPS faculty as an Associate Professor in 1990. After serving as Program Coordinator of the Fire and Safety Engineering Technology Program, Dr. Collins assumed the role of Department Chair for the Loss Prevention and Safety Department in 1998.

Dr. Collins' background includes serving as a design draftsman for a tank semi-trailer manufacturer. He has 24 years experience as a firefighter and has served in a large metropolitan fire department in northern Virginia and with a combination career/volunteer fire department in Uniontown, Pennsylvania. He has also been a local level fire instructor with the Pennsylvania Fire Academy.

As BioMarine Industry's first fire service safety specialist, Dr. Collins traveled North America conducting training sessions on closed circuit breathing apparatus and confined space monitoring instruments.

Dr. Collins holds an Associate of Science degree in Fire Science from Allegheny Community College of Pennsylvania, a B.S. in Industrial Arts Education from California University of Pennsylvania, and a Master of Education in Technology Education also from California University of Pennsylvania. Dr. Collins is nationally certified as a vocational carpentry instructor.

In July of 1993, Dr. Collins successfully defended his doctoral dissertation titled "Factors Which Influence the Implementation of Residential Fire Sprinklers," completing the requirements for the degree of Doctor of Education. Dr. Collins is a strong advocate for residential fire sprinklers. He believes this technology holds the real hope for changing fire death and injury statistics in the U.S.

Foreword

Disasters come in many forms. Natural disasters kill one million people around the world each decade, and leave millions more homeless. Natural disasters may include earthquakes, floods and flashfloods, landslides and mud flows, wild land fires, winter storms, and others. Technological disasters include house and building fires, hazardous materials, terrorism, and nuclear power plant emergencies. It is estimated that the economic damages from natural disasters have tripled in the past 30 years — rising from 40 billion dollars in the 1960s to 120 billion dollars in the 1980s. Some of the more recent natural disasters have by themselves caused billion dollar losses. For example, the World Health Organization has estimated that Hurricane Andrew in 1992 caused 30 billion dollars in damages. The Northridge, California earthquake in 1994 also caused approximately 30 billion dollars in damages. For other types of natural disasters such as flooding, it is estimated that the 1995 south central Alaska floods caused 10 million dollars in damages. The May 1995 Ft. Worth-Dallas storm left 16 dead and caused damages in excess of $900 million. Even worse were the 1995 southern California floods which left 11 dead and caused over 1.34 billion dollars in damages. The 1994 earthquake in southern California caused an estimated 13 to 20 billion dollars in damages.

Even more important is the issue of fires. Fire kills more Americans than all natural disasters combined. Each year more than 5000 people die in fires and over 25,000 are injured. It is estimated that the direct property loss exceeds 9 billion dollars. The U.S. has one of the highest fire death rates in the industrialized world. For example, in 1997, the U.S. fire death rate was 15.2 deaths per million population. Between 1993 and 1997, an average of 4500 Americans lost their lives, and another 26,500 were injured annually as the result of fire. Fire is the third leading cause of accidental death in the home and at least eighty percent of all fire deaths occur in residences.

The key to minimizing or controlling the cost and death toll of disasters is prevention. This is not to say that we can prevent natural disasters but we can minimize their effects. What this book shows is how we can evaluate, prepare for, react to, and minimize damage brought on by emergencies and disasters. One of the better examples of being prepared is a result of Executive Order 11988 of 1997. This is the floodplain management order is worded vigorously to reduce the risk of flood loss, minimize the impact of floods on human safety, health, and welfare, and restore and preserve the natural and beneficial values served by floodplains. The Office of Hydrology and U.S. Army Corps of Engineers have done an excellent job in working toward preventing the damage from flooding.

The Senate Environment and Public Works Committee recently approved unanimously an important piece of legislation to help the country to prevent disaster damage. Senate Bill 1691 authorizes the Federal Emergency Management Agency's Pre-Disaster Mitigation Initiative, Project Impact.

As this book discusses, the key is pre-planning and preparation — to develop an action plan and follow it through. People selection and training are crucial to the success of any disaster preparedness plan. This book will go through, step-by-step, what you need to do to prepare for disaster and prevent as much damage as possible.

Michael S. Schumann
Professor/Attorney
Department of Loss
Prevention and Safety
Eastern Kentucky University

Contents

Introduction

Chapter 1　Identifying the risks..1
　　What are the potential risks? 2
　　What is the probability of this risk happening? 2
　　Is the potential risk substantial? 2
　　If this risk becomes an event, what are the potential damages in
　　　terms of life, property, and other damages? 3

Chapter 2　Natural risks 5
　　Earthquakes 5
　　Volcanoes 7
　　　　Recent volcanic eruptions 8
　　Fires and explosions 9
　　U.S. hurricanes 11
　　Tornadoes 13
　　Major U.S. epidemics 14
　　Floods, avalanches, and tidal waves 14
　　Man and nature risks 15
　　　　Oil spills 15
　　　　Aircraft crashes 16
　　　　Dams 18
　　　　Shipwrecks 18
　　　　Mine explosions 18
　　　　Railroad accidents 19

Chapter 3　Emerging risks 21
　　Violence in the workplace 22
　　Cyberterrorism 25
　　Bioterrorism 26

Chapter 4　Governmental regulations 29
　　EPA (EPCRA) 30
　　OSHA standards 30

Chapter 5　Structural preparedness 33

Chapter 6 Coordinating with local assets 39

Chapter 7 Pre-planning for a disaster 43
   Internal actions 43
   External actions 46

Chapter 8 Eliminating, minimizing, and shifting risks 49

Chapter 9 Developing an action plan 53

Chapter 10 Developing the written plan 57

Chapter 11 Effective communication 61
   A universally accepted system of management and command 61
   Radios, telephones, and emergency operations centers 63
   Internal communications 64

Chapter 12 Selecting the right people 67
   Assembling an emergency response planning team 67
   Selecting employees to serve on the emergency response team
       or brigade 69

Chapter 13 Training for success 71

Chapter 14 Media control 83

Chapter 15 Shareholder factor 87

Chapter 16 After a disaster — minimizing the damage 91
   Critical stress debriefing 92
   Insurance companies 92
   Construction equipment 92
   Sale of debris 92
   Information to employees and families 93

Chapter 17 Governmental reactions 95
   De minimis violations 102
   Other or non-serious violations 103
   Serious violations 104
   Willful violations 106
   Repeat and failure to abate violations 107
   Failure to post violation notices 108
   Criminal liability and penalties 108

Chapter 18 Legal issues 115

Chapter 19 Disability issues	121
Title I — Employment provisions	134
Question 1: Who must comply with Title I of the ADA?	134
Question 2: Who is protected by Title I?	135
Question 3: What constitutes a disability?	136
A physical or mental impairment	136
Substantial limits	138
Major life activities	139
Question 4: Is the individual specifically excluded from protection under the ADA?	140
Title II — Public services	145
Title III — Public accommodations	147
Title IV — Telecommunications	148
Title V — Miscellaneous provisions	149
Chapter 20 Disaster preparedness assessments	151
Chapter 21 Personal disasters — use of criminal sanctions	155
Appendix A OSHA inspection checklist	171
Appendix B Employee workplace rights	173
Introduction	173
OSHA standards and workplace hazards	174
Right to know	175
Access to exposure and medical records	175
Cooperative efforts to reduce hazards	175
OSHA state consultation service	175
OSHA inspections	176
Employee representative	176
Helping the compliance officer	176
Observing monitoring	176
Reviewing OSHA Form 200	177
After an inspection	177
Challenging abatement period	177
Variances	177
Confidentiality	178
Review if no inspection is made	178
Discrimination for using rights	178
Employee responsibilities	180
Contacting NIOSH	181
Other sources of OSHA assistance	181
Safety and health management guidelines	181
Appendix C Web sites for disaster preparedness	183
Information and equipment	183

Appendix D Typical responsibilities 189
   Introduction 189
      Individuals 189
      Departments and agencies 189
      Non-governmental organizations 190
      Groups and teams 190
         Chief Executive Officer 190
         Emergency Program Manager 191
         EOC Manager 192
         Communications Coordinator 193
         Evacuation Coordinator 194
         Mass Care Coordinator 194
         Mass Care Facility Manager 195
         Public Health Coordinator 196
         County Coroner/Medical Examiner 197
         Emergency Medical Services 198
         Mental Health Agencies 198
         Health and Medical Facilities 199
         Public Information Officer 199
         Resource Manager 200
         Needs Group 201
            Needs Analyst 201
            Needs Liaisons 201
         Supply Group 201
            Supply Coordinator 202
            Donations Team 202
            Procurement Team 202
            Personnel Team 203
            Financial Officer 203
            Legal Advisor 203
         Warning Coordinator 203
         Agricultural Extension Agent 204
         Fire Department 204
         Police Department 204
         Public Works 205
         Education Department (Superintendent of Education) 206
         Legal Department 207
         Military Department 207
         Animal Care and Control Agency 207
         Comptroller's Office (or equivalent) 209
         Department of General Services (or equivalent) 209
         Office of Personnel, Job Service 210
         Office of Economic Planning (or equivalent) 210
         Department of Transportation (or equivalent) 210
         Private Utility Companies 210
         EAS Stations 210

   Local Media Organizations 210
   Volunteer Organizations 210
   American Red Cross (local) 211
   Salvation Army (local) 211
   Non-profit Public Service Organizations 211
   Communications Section Team Members 211
   Distribution Group 211
     Distribution Coordinator 212

Appendix E **Potential sources of disaster preparedness and management assistance through local colleges and universities** 213

Index 243

Introduction

> "There are no disasters in business that you can't avoid — if you see them coming and make the adjustments."
>
> T. Boone Pickens, Jr.

> "The crisis of yesterday is the joke of tomorrow."
>
> H.G. Wells

By simply watching the evening news or reading a newspaper, we quickly find that disasters of various types happen to individuals, companies, and countries on virtually a daily basis throughout the world. Disasters take various forms ranging from natural disasters, such as tornados, to man-made disasters, such as workplace violence incidents, and happen on a far too frequent basis. No matter what type of disaster befalls the individual, organization or country, the results are typically the same, i.e., substantial loss of life, money, assets, and productivity.

Today's risks of disasters have evolved substantially to include areas far beyond the natural disasters of the past. Disasters now encompass areas such as cyberterrorism, product tampering, biological threats, and ecological terrorism which were virtually unheard of a few short years ago. The results of disaster risks can be just as devastating to an organization as a natural disaster; however, the prevention and proactive measures taken are substantially different. Today's safety professional is faced with a myriad of new and different reactions and issues ranging from control of the media to shareholder reaction, which were not given consideration in the disaster preparedness programs of the past. The world is changing, technology is changing, risks are increasing and evolving. The safety profession must adapt in order to prevent potential disasters from happening where possible, minimize the risks where prevention is not possible, and appropriately react to keep the damages to a minimum.

Appropriate planning and preparedness **before** a disaster happens are essential to minimizing the risks and the resulting damages. Individuals involved in the disaster preparedness efforts must be appropriately selected and trained in order that their reaction and decision-making during a crisis are appropriate. Risks that cannot be appropriately managed internally should be assessed for external protection such as with insurance. Reaction after the disaster must be sure and coordinated in order to minimize the disaster damage as well as avoid causing further damage to the remaining assets. Intangible damage control must to tried and tested in order to avoid long term damage to efficacy assets.

In this text, we have designed a new and innovative method through which to prepare companies and organizations to address the substantial risk of disasters in the workplace. Our methodology encompasses not only tried and true proactive methodology utilized by safety professionals for decades in addressing natural disasters, but also addresses the often overlooked areas during the reactive and post-disaster phases. Society's progress in technology combined with new terroristic activities have required safety professionals to rethink the standard *modus operandi* in the area of disaster preparedness and expand their proactive and reactive measures to safeguard their company's or organization's assets on all levels and at all times. One disaster can decimate all of the years of effort, creativity and sweat, as well as monetary and physical assets of a company or organization in virtually the blink of an eye. Safety professionals are charged with the responsibility of identifying these risks, preparing to protect against these risks, and reacting properly if these risks should develop. We hope this text will assist you in opening your company's or organization's eyes and prepare you to appropriately address this important area of workplace safety.

Remember, as Lee Whistler said, "Few people plan to fail, they just fail to plan." Disaster can happen and has happened to many individuals and companies on a daily basis. Preparedness before a disaster situation is key to minimizing the potential risks and damages which can occur in terms of human, property, and efficacy losses. After a disaster happens, it is too late.

chapter one

Identifying the risks

> "Dis-as-ter: a calamitous event, especially one occurring suddenly and causing great loss of life, damage, or hardship, as a flood, airplane crash, or business failure."
>
> Blacks Law Dictionary

> "Nations have passed away and left no traces, Any history gives the naked cause of it — one single simple reason in all cases; they fell because their people were not fit."
>
> Rudyard Kipling

Risks of varying types and magnitudes exist in every workplace on a daily basis. However, some risks are far greater and can be disastrous if not identified and properly addressed in terms of reducing the probability of the risk, where feasible, protecting assets through shifting all or a percentage of the risk, and minimizing the potential harm of the risk in a disaster event.

The initial step in any disaster preparedness endeavor is to identify the potential risks, assess their viability, evaluate the probability of risks occurring, and appraise the potential damage. Potential risks will vary from operation to operation, facility to facility, and location to location. Thus, it is important that safety professionals properly identify and assess each facility and operation on an individual basis and customize the preparedness program to meet the needs of the each facility. **There is no one basic emergency and disaster plan that fits all facilities and operations.**

The initial step in identifying the potential risks to an operation or facility is to assess the location, the process, the environment, the structure, and related factors. Below is a basic assessment list to assist you with this initial assessment:

What are the potential risks?

- Is the operation or facility located near any natural risks? Examples include the following:
  - Facility located on an earthquake fault
  - Facility located near a volcano
  - Facility located in a hurricane zone
  - Facility located in a heavy snow area
  - Facility located on or near a river
  - Facility located near a forest fire area
- Is the operation or facility located near other operations or facilities possessing potential risks that could be transferred? Examples include the following:
  - Facility located next to a petrochemical refinery
  - Facility located near a high explosives manufacturing facility
  - Facility located near a military air field
  - Facility located near a nuclear facility

What is the probability of this risk happening?

Upon completion of the listing of all potential risks, assess each potential risk as to the probability of an incident or event occurring at the facility. The probability can be assessed utilizing a numerical formula or other assessment methodology; however, there is always a certain amount of subjective assessment which is included in the measurement. For example, if the facility is located in a desert, there is minimal potential of flooding at the facility. Conversely, if the facility is located on the slopes of a dormant volcano, it has a higher probability of volcanic activity than the facility located in the desert.

Is the potential risk substantial?

An assessment identifies which potential risks are greatest and which potential risks are substantially lower in terms of dollar losses, potential of injury or death, and other potential losses. If the risk is substantial, such as the potential of a fire in a paper facility, then appropriate resources can be expended to develop appropriate safeguards and the risk of loss can be shifted through insurance. However, where the potential risk is identified as being low, then an assessment must be made of the time, resources, and manpower necessary to minimize this potential risk. For example, the potential of a fire in the wooden hay storage buildings located one mile from the main facility is substantially high. However, the potential loss in terms of life, property, and structure is substantially low. Should the professional include the hay storage structures within the emergency and disaster planning?

Should insurance be purchased to cover the risk of loss of the hay facility? Can the organization risk losing the hay facility to fire without a detrimental effect on the operation or profitability of the organization? Is this a risk that the organization can assume without expending resources?

If this risk becomes an event, what are the potential damages in terms of life, property, and other damages?

After identifying all of the potential risks and assessing their probability, safety professionals should analyze and estimate the potential damages to life, property, productivity, and efficacy in monetary terms. Although this type of analysis will not be exact, it can provide your management team with an estimate through which to base their decision-making as to the dollars and manpower provided to the prevention efforts as well as acquisition of insurance or other protections. This type of analysis often requires *worst-case scenario* assessments with various *what if* assessments. This is one of the most difficult areas because there are no hard and fast rules for safety professionals to follow in order to know how narrow or how broad to make assessments.

For example, the worst-case scenario for the facility located on the slopes of a volcano is that the volcano erupts suddenly and without warning. In this scenario, all personnel, the complete facility, all products, and all continuing operations are lost. How much would this type of disaster cost the company? How much money is an employee killed on the job determined to be worth in the specific state? How much is the facility worth? How much is the equipment in the facility worth? How much production would be lost while a new facility is located or the product line is shifted to another company operation? What are the efficacy losses to the company? What losses are covered by insurance? What losses would not be covered by insurance? How would the company function with the loss of several key management team members?

Basic Assessment List

Potential Loss	Items	Potential Loss ($)	Insurance ($)	Loss ($)
Facility	50,000 sq. ft.	$2,000,000	$2,000,000 with 50,000 deductible	$50,000
Equipment	List equipment	$1,000,000	$1,000,000 with 50,000 deductible	$50,000
10 employees	Workers Comp. (* Potential of other legal actions and costs)	$1,000,000	Self-insured	$1,000,000

* List other potential losses and estimated costs.

Upon completion of a risk analysis, the safety professional would possess a workable assessment tool containing the potential risks, the probability of the risk occurring as well as a monetary assessment through which he/she can address the management team to acquire the necessary resources to reduce or eliminate the potential risks through a proactive, preventative plan or action. Additionally, this type of thorough analysis will provide the safety professional with a method through which to educate his/her management team to identify other related potential risks in the operation to avoid potential risks in the future.

In essence, this type of analysis and assessment *opens the eyes* of many management team members who become narrowly focused on the specific day-to-day operations of the business. Safety professionals will often need to educate management team members and broaden their thinking in order to acquire the necessary resources to properly develop a proactive plan of action to address potential catastrophic risk in the workplace. Emergency and disaster planning is not cheap and there is no instantaneous return on investment for the management team. However, through proper planning and appropriate management of these potential risks, the potential of a disaster in the workplace can be minimized, and if a disastrous event does occur, the losses, and thus the costs, will be minimized. Emergency and disaster preparedness pays enormous dividends when disaster strikes, but like the safety and loss prevention function, does not make anything or generate any income for the company on a weekly or monthly basis. Emergency and disaster preparedness is not unlike insurance delete — it's better to pay your premiums beforehand in hopes that you never have an accident, but when you do have an accident, you definitely want this protection. When disaster strikes, as it does daily somewhere in the world, you definitely want to be prepared.

chapter two

Natural risks

> "Adapt or perish, now as ever, is nature's inexorable imperative."
>
> H.G. Wells

> "Nature is a catchment of sorrows."
>
> Maxine Kumin

In assessing the potential risks of an operation or facility, the initial risks most safety professionals address are the naturally occurring risks which we live with as human beings inhabiting the planet earth, namely natural disaster risks. The potential risks will vary depending on your location on the planet as well as the types of operations being performed and the facility structure. For example, if your company was operating a facility near a volcano, the most devastating potential risk would be the volcano erupting. However, for most operations, the potential risk of a volcanic eruption is minimal because the nearest volcano is hundreds, if not thousands of miles from the facility. Each of these facilities possesses the potential risk of a volcanic eruption; however, the probability of a disaster resulting from a volcanic eruption is far greater in the first facility than it is at the second facility.

Natural risks are inherent but are often overlooked or taken for granted when assessing potential risks. Safety professionals often acquire microscopic vision due to the extended focus on the specific hazards of the workplace. However, safety professionals should expand their field of vision when assessing the potential risks of their facility or operations to include proper analysis and assessment of the potential natural risks at their location. Below is a synopsis of the various potential risks as well as a brief historical view of these potential risks.

Earthquakes

Earthquake: any abrupt disturbance within the earth that is tectonic or volcanic in origin and that results in the generation of elastic waves. The passage of such seismic waves through the earth often causes violent shaking at its surface.

The origin and distribution of most major earthquakes can be explained using the plate tectonics theory. This theory postulates that the earth's surface is made up of a number of large, rigid plates that move in relation to one another and interact at their boundaries. The severest earthquakes tend to occur at convergent plate boundaries where one plate descends beneath the other. Most of these quakes originate more than 300 km (190 miles) below the surface and are associated with island arcs and trenches. Seismicity occurs near the margins where plates separate or slide past one another. Quakes at such sites tend to be of lower magnitude and are fairly shallow. In all such boundary regions, seismic waves are generated by the sudden fracturing of rock, which results when elastic strain accumulated during tectonic processes exceeds the strength of the rock.

Three major zones of seismicity have been identified: (1) the circum-Pacific belt, which lies along plate margins around the Pacific Ocean and includes the well-known seismically active areas of Japan, Indonesia, New Guinea, the Andes Mountains, the western part of Central America, and the San Andreas Fault region of California, (2) the trans-Asiatic belt, extending from Mediterranean Europe eastward through Asia to the Pacific, and (3) the mid-ocean ridges, which form a connected worldwide rift system.

Some earthquakes occur outside of these belts, away from plate boundaries. These intraplate earthquakes must be explained by mechanisms other than plate motions and suggest that stresses occasionally can exceed the strength of rock masses even within plates.

Some phenomena akin to earthquakes, usually relatively minor, have been triggered by human activities that disturb the equilibrium of subsurface rock layers, e.g., underground nuclear testing, impounding of water behind high dams, and the pumping of liquid waste into the earth through deep wells.

The location of an earthquake is determined with a seismograph. This instrument records the oscillation of the ground caused by seismic waves that travel from their point of origin through the earth or along its surface. A seismogram of a nearby earthquake is fairly simple, showing the arrival of P (or primary) waves, those that vibrate in the direction of propagation; slower-traveling S (or secondary) waves, those that vibrate at right angles to the direction of propagation; and surface waves, those of extremely high amplitude that skirt along the earth's surface. In the case of distant earthquakes, the seismogram pattern tends to be more complex because it shows various types of seismic waves that originate from one point but are then reflected or refracted within the earth's crust before reaching the seismograph. The relation between the arrival time of these waves and the epicentral distance (i.e., the distance from the point of origin) is expressed by a time-distance curve in which the arrival time is read on the vertical axis and the epicentral distance on the horizontal axis. If the arrival times of various seismic waves are read on the seismogram at a recording station and compared with standard time-distance curves, then the distance to the center of an earthquake can be ascertained.

The magnitude of an earthquake is usually expressed in terms of a logarithmic scale based on seismograph recordings of seismic-wave amplitudes. The numerical scale is so arranged that each increase in magnitude of one unit represents a 10-fold increase in earthquake size, i.e., an earthquake of magnitude 8 is 10,000 times as large as one of magnitude 4. Whereas the latter would be capable of causing only slight damage, the former constitutes a devastating seismic event. The scales that are commonly used are derivations of the Richter scale, which was introduced in southern California in 1935 and which, with successive refinements, held currency among seismologists for more than 40 years.

The magnitude of an earthquake differs from its intensity, which is the perceptible degree of shaking of the earth's surface and the attendant damage at any given location. In general, a quake's intensity decreases with distance from its epicenter, but other factors, including surface geology, may have a significant bearing on its effects on man-made structures. Large earthquakes have caused some of the worst disasters in history. No other natural phenomenon is as destructive over so large an area in so short a time. A major earthquake that struck Shensi province of China in 1556, for example, is estimated to have killed nearly 830,000 people while destroying entire towns and villages. The violent motions of the surface during large quakes can topple buildings. People are crushed and buried under the collapsing structures or are burned to death in ensuing building fires. Destructive, too, are the landslides and mudslides that may accompany an earthquake, as are tsunamis, the huge seismic sea waves induced by a disturbance in the adjacent seabed or by a submarine landslide triggered by an earthquake.

Much research has been devoted to earthquake prediction since the mid-1960s, most notably by seismologists in China, Japan, Russia, and the U.S. Various advances notwithstanding, no method has yet been devised to predict the time, place, or magnitude of earthquakes with a high degree of accuracy or consistency. Seismologists have found that major earthquakes are often preceded by certain measurable physical changes in the environment around their epicenters. These so-called precursor phenomena include the degree of crustal deformation in fault zones, occurrence of dilatancy (i.e., an increase in volume) of rocks, and a rise in radon concentrations in wells. Continual monitoring and close scrutiny of these and other related phenomena are expected to improve prediction capability in the future.

Volcanoes

Volcano: any vent in the crust of the earth or other planet or satellite (e.g., Jupiter's Io) from which molten rock, pyroclastic debris, and steam issue.

Volcanoes are commonly divided into two broad types: fissure and central. Each type is associated with a different mode of eruption and surface structure.

Fissure volcanoes are much more common than those of the central type. They occur along fractures in the crust and may extend for many kilometers. Lava, usually of basaltic composition, is ejected relatively quietly and continuously from the fissures and forms enormous plains or plateaus of volcanic rock. Submarine fissure eruptions are common along the crests of mid-ocean ridges and are pivotal in seafloor spreading. When molten rock is extruded under water, pillow lava (piles of sack-shaped rock masses measuring up to several meters in diameter) are often formed.

Central volcanoes have a single vertical lava pipe and tend to develop a conical profile. The volcanic cone is generally built up of a succession of lavas, ignimbrites, and welded tuffs (porous rock formed by the cementation of solidified volcanic ash and dust particles). Lava flows from the throat of a central volcano following the easiest path downhill, its flow pattern strongly influenced by the topography.

The shape of any given volcanic landform depends on a variety of local circumstances and on the relative abundances of lavas, tuffs, and ignimbrites. This, in turn, depends on the composition of the magma arriving at the surface. The lower the viscosity, the more readily the lava flows away from the throat or fissure. There is, as a consequence, relatively little tendency to build up a steep-sided cone. The more viscous the magma, the greater the tendency to chill and solidify close to the source and to form a cone. In many cases, highly viscous lava also may clog the throat of the volcano, causing a pressure buildup that can only be relieved by violent explosions and nuées ardentes. Such eruptions, exemplified by those of Mount St. Helens in southwestern Washington state during the early 1980s and Vesuvius in AD 79, may completely remove the top of a volcanic cone and occasionally part of the interior of the cone as well. The resultant roughly circular hollow is called a *caldera*. Further eruption may lead to the formation of a lava lake within the cone, and if the lava cools and solidifies, the inward drainage of rainwater may produce a water lake on the surface of the lava lake. A caldera may also form without an explosion by the collapse of the top of the cone into an underlying accumulation of magma. Kilauea on southeastern Hawaii is an excellent example of a large volcanic cone with a well-developed caldera produced by collapse.

Recent volcanic eruptions

March 10, 1933 — Long Beach, CA: 117 left dead by earthquake.
May 30, 1935 — Pakistan: earthquake at Quetta killed 30,000–60,000 people
January 24, 1939 — Chile: earthquake razed 50,000 sq. mi.; 30,000 killed
August 15, 1950 — India: earthquake affected 30,000 sq. mi. in Assam; 20,000–30,000 people believed killed
March 28, 1964 — Alaska: strongest earthquake ever to strike North America hit 80 miles east of Anchorage; followed by seismic wave 50 feet high that traveled 8445 miles at 450 miles per hour; 117 killed

Chapter two: Natural risks

May 31, 1970 — Peru: earthquake left more than 50,000 dead, 17,000 missing

December 22, 1972 — Managua, Nicaragua: earthquake devastated city, leaving up to 6000 dead

February 4, 1976 — Guatemala: earthquake left over 23,000 dead

July 28, 1976 — Tangshan, China: earthquake devastated 20 sq. mi. area of city, leaving 242,000 confirmed dead, with estimates placing toll as high as 655,000

August 17, 1976 — Mindanao, Philippines: earthquake and tidal wave left up to 8000 dead or missing

September 16, 1978 — Tabas, Iran: earthquake destroyed city in eastern Iran, leaving 25,000 dead

September 19, 1985 — Mexico: earthquake registering 8.1 on Richter scale struck central and southwestern regions, devastating part of Mexico City and three coastal states; an estimated 25,000 killed

December 7, 1988 — Armenia: earthquake measuring 6.9 on the Richter scale killed nearly 25,000, injured 15,000, and left at least 400,000 homeless

October 17, 1989 — San Francisco Bay Area: earthquake measuring 7.1 on Richter scale killed 67 and injured over 3000; over 100,000 buildings damaged or destroyed; damage cost city billions of dollars

June 21, 1990 — Northwestern Iran: earthquake measuring 7.7 on Richter scale destroyed cities and villages in Caspian Sea area; at least 50,000 dead, over 60,000 injured, and 400,000 homeless

January 17, 1994 — San Fernando Valley, CA: earthquake measuring 6.6 on Richter scale killed 61 and injured over 8000; damage estimated at $20 billion

January 17, 1995 — Osaka, Kyoto and Kobe, Japan: 5100 killed and 26,800 injured; estimated damage $100 billion

May 30, 1998 — Northern Afghanistan: magnitude 7.1 earthquake and aftershocks killed an estimated 5000 and injured at least 1500 (A February 4th quake in same area killed about 2300)

January 25, 1999 — Western Colombia: 1124 dead and 4000 injured in magnitude 6 earthquake in and around the city of Armenia; more than 200,000 left homeless; city plagued by looting and violence after disaster

These are just a few of the most recent earthquake related disasters. If your facility or operations are located in an earthquake prone area, this risk cannot be minimized.

Fires and explosions

Fire: the rapid burning of combustible material with the evolution of heat and usually accompanied by flame. It is one of man's essential tools, control of which helped propel him forward on the path of civilization.

The original source of fire undoubtedly was lightning, and such fortuitously ignited blazes remained the only source of fire for eons. For some years, Peking man (about 500,000 B.C.) was believed to be the earliest user of fire. Evidence uncovered in Kenya in 1981 and in South Africa in 1988, however, suggests that the earliest controlled use of fire by hominids dates from about 1,420,000 years ago. Not until about 7000 B.C. did Neolithic man acquire reliable fire-making techniques, in the form of drills, saws, other friction-producing implements, or of flint struck against pyrites. Even then it was more convenient to keep a fire alive permanently than to reignite it.

The first human beings to control fire gradually learned its many uses. Not only did they use fire to keep warm and cook their food, but they also learned to use it in fire drives in hunting or warfare, to kill insects, to obtain berries, and to clear forests of underbrush so that game could be better seen and hunted. Eventually they learned that the burning of brush produced better grasslands and therefore more game.

With the development of agriculture in Neolithic times in the Middle East about 7000 B.C., there came a new urgency to clear brush and trees. The first agriculturists made use of fire to clear fields and to produce ash to serve as fertilizer. This practice, called slash-and-burn cultivation, persists in many tropical areas and some temperate zones today.

The step from the control of fire to its manufacture is great and took hundreds of thousands of years. Not until Neolithic times is there evidence that human beings actually knew how to produce fire. Whether a chance spark from striking flint against pyrites or a spark made by friction while drilling a hole in wood gave human beings the idea for producing fire is not known, but flint and pyrites, as well as fire drills, have been recovered from Neolithic sites in Europe.

Most widespread among prehistoric and later primitive peoples is the friction method of producing fire. The simple fire drill, a pointed stick of hard wood twirled between the palms and pressed into a hole on the edge of a stick of softer wood, is almost universal. The fire plow and the fire saw are variations on the friction method common in Oceania, Australia, and Indonesia. Mechanical fire drills were developed by the Eskimo, ancient Egyptians, Asian peoples, and a few American natives. A fire piston that produced heat and fire by the compression of air in a small tube of bamboo was a complex device invented and used in southeastern Asia, Indonesia, and the Philippines. About 1800, a metal fire piston was independently invented in Europe. In 1827 the English chemist John Walker invented the friction match containing phosphorous sulfate, essentially the same match which is used today.

Why do we provide this history of fire? Because uncontrolled fire remains one of the major causes of death and property damage in today's society. Proper preparedness in the event of a fire in your operations or facilities is essential to safeguard life and minimize property damage. Some of the more recent fires and explosions include the following:

Chapter two: Natural risks 11

> November 28, 1942 — Boston, MA: Coconut Grove nightclub fire killed 491; primary event leading to new fire related laws and regulations
> July 6, 1944 — Hartford, CT: fire and ensuing stampede in main tent of Ringling Brothers Circus killed 168, injured 487
> December 7, 1946 — Atlanta, GA: fire in Winecoff Hotel killed 119
> April 16–18, 1947 — Texas City, TX: most of the city destroyed by a fire and subsequent explosion on the French freighter *Grandcamp* carrying a cargo of ammonium nitrate. At least 516 were killed and over 3000 injured
> May 22, 1967 — Brussels: fire in L'Innovation, major department store, left 322 dead
> November 29, 1973 — Kumamoto, Japan: fire in Taiyo department store killed 101
> February 1, 1974 — Sao Paulo, Brazil: fire in upper stories of bank building killed 189 persons, many of whom leaped to their deaths
> May 28, 1977 — Southgate, KY: fire in Beverly Hills Supper Club; 167 dead; event leading to new fire related regulations
> December 18–21, 1982 — Caracas, Venezuela: power/plant fire left 128 dead
> December 31, 1986 — San Juan, PR: arson fire in Dupont Plaza Hotel was set by three employees, killing 96 people
> October 23, 1989 — Pasadena, TX: a huge explosion followed by a series of others and a raging fire at a plastics manufacturing plant owned by Phillips Petroleum Co. killed 22 and injured more than 80 persons; a large leak of ethylene was presumed to be the cause
> March 25, 1990 — New York City: arson fire in the illegal Happy Land Social Club, in the Bronx, killed 87 people
> May 10, 1993 — near Bangkok, Thailand: fire in doll factory killed at least 187 persons and injured 500 others; world's deadliest factory fire
> March 24, 1999 — Chamonix, France: Belgian truck carrying margarine and flour broke out into flames in the MontBlanc tunnel, trapping dozens of cars; death toll was at least 42

As can be seen, fires do still happen on a frequent basis in industry and in other arenas. Although fire protection activities are often perceived as routine, preparedness for fire related disasters remains essential in American industry.

U.S. hurricanes

Hurricane: a tropical cyclone formed over the North Atlantic, E. North Pacific, W. South Pacific, and Indian oceans in which the winds attain speeds greater than 75 mph (121 km/hr). A tropical cyclone passes through two stages, tropical depression and tropical storm, before reaching hurricane force. An average of 3.5 tropical storms per year become hurricanes; one to three of these approach the U.S. coast. Hurricanes usually develop between July and October. A hurricane is nearly circular in shape, and its winds cover an area about 500 mi (800 km) in diameter. As a result of the extremely low central air pressure (around 28.35 in/72 cm of mercury), air spirals inward toward the hurricane's eye, an almost calm area about 20 mi (30 km) in diameter. Hurricanes, which may last from 1 to 30 days, usually move westward in their early stages and then curve northward toward the pole. Deriving their energy from warm tropical ocean water, hurricanes weaken after prolonged contact with colder northern ocean waters, becoming extratropical cyclones; they decay rapidly after moving over land areas. The high winds, coastal flooding, and torrential rains associated with a hurricane may cause enormous damage. Tropical cyclones that form over the E. North Pacific Ocean and its seas are called typhoons; those over the Indian Ocean and its seas, cyclones.

Hurricanes can cause extreme damage and loss of life if proper preparedness is not taken to safeguard assets. Some of the most recent hurricanes in the U.S. include the following:

August 29–September 13, 1960 — Florida to New England: "Donna" killed 50 in the U.S., 115 deaths in Antilles — mostly from flash floods in Puerto Rico

September 3–15, 1961 — Texas coast: "Carla" devastated Texas gulf cities, taking 46 lives

August 27–September 12, 1965 — Southern Florida and Louisiana: "Betsy" killed 75 people

August 14–22, 1969 — Mississippi, Louisiana, Alabama, Virginia, and West Virginia: 256 killed and 68 persons missing as a result of "Camille"

October–November, 1985 — Louisiana and the Southeast: though only a category 1 hurricane, "Juan" caused severe flooding and $1.5 billion in damages; 63 lives were lost

September 10–22, 1989 — Caribbean Sea, South Carolina, and North Carolina: "Hugo" claimed 49 U.S. lives (71 killed overall); $4.2 billion paid in insurance claims

August 22–26, 1992 — South Florida, Louisiana, and Bahamas: Gulf Coast hurricane "Andrew," with damage estimated at $15–$20 billion, is most costly hurricane in U.S. history

September 5, 1996 — North Carolina and Virginia: "Fran," a category 3 hurricane, took 37 lives and caused $5 billion in damage

If your operations or facilities are located in a hurricane prone area, proper preparedness is essential. Remember, preparedness for a hurricane is significantly different than the usual preparedness efforts. Evacuation is usually inside rather than outside of the facility and preparation duration is usually longer.

Tornadoes

Tornado: a violent cyclonic storm, relatively small in diameter but with rapidly rotating winds that form a funnel cloud, or vortex, and moves over land. Whirlwinds are similar but smaller storms of less intensity. When a tornado forms or passes over a water surface, it is called a waterspout.

The name tornado comes from the Spanish *tronada* (thunderstorm), which supposedly was derived from the Latin *tornare* (to make round by turning). The most violent of atmospheric storms, a tornado is a powerful vortex, or twister, whose rotational speeds are estimated to be close to 480 kilometers per hour, but may occasionally exceed 800 kilometers per hour. The direction of rotation in the Northern Hemisphere is usually, though not exclusively, counterclockwise.

The first visible indication of tornado development is usually a funnel cloud, which extends downward from the cumulonimbus cloud of a severe thunderstorm. As this funnel dips earthward, it becomes darker because of the debris forced into its intensifying vortex. Some tornadoes give no visible warning until their destruction strikes the unsuspecting victims. Tornadoes often occur in groups, and several twisters sometimes descend from the same cloud base.

The forward speed of an individual tornado is normally 48 to 64 kilometers per hour but may range from nearly zero to 112 kilometers per hour. The direction of motion is usually from the southwest to the northeast, although tornadoes associated with hurricanes may move from the east. The paths of twisters average only several hundred meters in width and 26 kilometers in length, but large deviations from these averages may be expected — e.g., a devastating tornado that killed 689 persons in Missouri, Illinois, and Indiana in the midwestern United States on March 18, 1925, was, at times, 1.6 kilometers wide; its path extended 352 kilometers.

In the short time that it takes to pass, a tornado causes vast destruction. There have been cases reported in which blades of straw were embedded in fence posts; a schoolhouse with 85 pupils inside was demolished, and the pupils were carried 137 meters with none killed; five railway coaches, each weighing 70 tons, were lifted from their tracks; one coach was moved 24 meters.

Although much remains to be learned about tornado formation and movement, advances have been made in the effectiveness of tornado detection and warning systems. These systems involve analyses of surface and upper-air weather, detection and tracking of atmosphere changes by radar, and the spotting of severe local storms.

Although tornadoes are more prevalent in the *tornado belt* of the Midwest, tornadoes can happen anywhere. Some of the most recent tornadoes causing loss of life and significant damage include the following:

May 11, 1953 — Waco, TX: a single tornado struck, killing 114
June 8, 1953 — Flint, MI: tornado killed 116

June 9, 1953 — Worcester, MA: tornado hit town, causing 90 deaths
May 25, 1955 — Udall, KS: tornado killed 80
April 11, 1965 — Tornadoes in Iowa, Illinois, Indiana, Ohio, Michigan, and Wisconsin caused 256 deaths
April 3–4, 1974 — East, South, and Midwest: a series of 148 twisters comprised the deadly 1974 *Super Tornado Outbreak* that struck 13 states; 330 died and over 5000 were injured in a damage path covering more than 2500 miles; it was the worst tornado outbreak in U.S. history
January 17–22, 1999 — Tennessee and Arkansas: a series of tornadoes swept through, leaving 17 dead; damages were estimated at $1.3 billion
May 3, 1999 — Oklahoma: unusually large twister, thought to have been a mile wide at times, killed 41 people and injured at least 748 others in Oklahoma; a separate tornado killed another 5 and injured about 150 in Kansas; damages totaled at least $1 billion

Major U.S. epidemics

With the new OSHA standards for such communicable diseases as tuberculosis and the potential impact on the workforce, prudent employers may want to include preparedness action to address the potential of epidemics in the workplace. Some of the most recent epidemics to be considered include the following:

1916 — Nationwide: over 7000 deaths occurred and 27,363 cases were reported of polio (infantile paralysis) in America's worst polio epidemic
1918 — March–November, Nationwide: outbreak of Spanish influenza killed over 500,000 people in the worst single U.S. epidemic
1949 — Nationwide: 2720 deaths occurred from polio; 42,173 cases were reported
1952 — Nationwide: polio killed 3300; 57,628 cases reported; worst epidemic since 1916
1981 to December 1998 — total U.S. AIDS cases reported to Centers for Disease Control: 688,200

Although epidemics are usually a larger national problem, prudent employers may want to consider preparedness actions that include regional epidemics and/or site-specific outbreaks.

Floods, avalanches, and tidal waves

Depending on the location of the operation, floods, tidal waves, and avalanches may pose a substantial risk of harm. For example, if your operation is located on a flood plain, the risk of flood is substantially higher than if your operation is located on a mountain. However, the risk of avalanche may be substantially higher at this location when compared to the potential of flooding. All potential risks should be assessed and analyzed. Some of the more recent disasters in these areas include the following:

January 18–26, 1969 — Southern California: floods and mudslides from heavy rains caused widespread property damage; at least 100 dead; another downpour (February 23–26) caused further floods and mudslides with at least 18 dead

February 26, 1972 — Man, WV: more than 118 died when slag-pile dam collapsed under pressure of torrential rains and flooded 17-mile valley

June–August 1993 — Illinois, Iowa, Kansas, Kentucky, Minnesota, Missouri, Nebraska, North Dakota, South Dakota, Wisconsin: two months of heavy rain caused Mississippi River and tributaries to flood in 10 states, causing almost 50 deaths and about $12 billion in damage to property and agriculture in Midwest; almost 70,000 left homeless

December 1996–January 1997 — U.S. West Coast: torrential rains and snowmelt produced severe floods in parts of California, Oregon, Washington, Idaho, Nevada, and Montana, causing 36 deaths and about $2–3 billion in damage

March 1997 — Ohio and Mississippi Valleys: flooding and tornadoes plagued Arkansas, Missouri, Mississippi, Tennessee, Illinois, Indiana, Kentucky, Ohio, and West Virginia; 67 were killed and damage totaled approximately $1 billion

July 17, 1998 — New Guinea: spurred by an undersea earthquake, three tsunamis wiped out entire villages in the northwestern province of Sepik; one tidal wave reported by survivor to be 30 ft. high; at least 2000 found or presumed dead; many who were injured by the tsunamis were later killed by gangrene infections

Summer 1999 — Asia: flooding plagued Asia again after weeks of torrential downpours; more than 950 killed and millions left homeless in S. Korea, China, Japan, the Philippines, and Thailand

Man and nature risks

In recent years, the interaction between man's technology and nature has created new potential risks that should be considered in the overall risk assessment. If your operation or facility is utilizing technology that possesses the potential of substantial risk and/or if your operation is located near another operation or facility possessing substantial risk factors that could encompass your operation or facility in the event of an event, these potential risks should be analyzed:

Oil spills

Oil or petroleum spills are most often caused by human failure, but possess a major impact on the environment as well as operations and facilities located in or near a spill area. Major spills can cause major business interruption as well as long term damage to an operation dependent upon the natural resources such as salmon fishing, eco-tourism, or water bottling. Some of the major oil spills include the following:

> March 16, 1978 — Portsall, France: wrecked supertanker *Amoco Cadiz* spilled 68 million gallons, causing widespread environmental damage over 100 miles of Brittany coast; world's largest tanker disaster
> June 3, 1979 — Gulf of Mexico: exploratory oil well *Ixtoc 1* blew out, spilling an estimated 140 million gallons of crude oil into the open sea; although it is the largest known oil spill, it had a low environmental impact
> March 24, 1989 — Prince William Sound, AK: tanker *Exxon Valdez* hit an undersea reef and spilled 10 million plus gallons of oil into the waters, causing the worst oil spill in U.S. history
> December 19, 1989 — Las Palmas, Canary Islands: explosion in Iranian supertanker, the *Kharg-5*, tore through its hull and caused 19 million gallons of crude oil to spill out into the Atlantic Ocean about 400 miles north of Las Palmas, forming a 100-square-mile oil slick
> January 25, 1991 — Southern Kuwait: during the Persian Gulf War, Iraq deliberately released an estimated 460 million gallons of crude oil into the Persian Gulf from tankers at Minaal-Ahmadi and Sea Island Terminal 10 miles off Kuwait; spill had little military significance; on January 27, U.S. warplanes bombed pipe systems to stop the flow of oil
> August 12, 1994 — Ursinsk, Russia: huge oil spill from ruptured pipeline
> September 8, 1994 — Russia: a dam built to contain oil burst and spilled oil into Kolva River tributary; U.S. Energy Department estimated spill at 2 million barrels; Russian state-owned oil company claimed spill was only 102,000 barrels
> February 15, 1996 — Welsh coast: supertanker *Sea Empress* ran aground at port of Milford Haven, Wales, spewed 70,000 tons of crude oil, and created a 25-mile slick

Disaster situations created by others that impact your operations or facilities is often a fact of life in today's industrial society. Careful consideration should be provided to the potential risks that could be created by other companies and operations within the vicinity of your operations.

Aircraft crashes

With today's global economy, air travel has become commonplace in the work schedule of many executives and members of the management team. Key individuals within your company may be flying via every level of the aviation spectrum ranging from corporate jets to helicopters to small commuter airlines. The risk of potential loss can occur during domestic or international travel. To minimize a potential loss which could be devastating to the organization, many companies require key executives to fly separately

Chapter two: Natural risks 17

to locations. Additionally, facilities and operations located near aviation centers may want to consider the potential of aircraft making contact with towers and facilities. Some of the incidents to consider include the following:

> August 3, 1975 — Agadir, Morocco: Chartered Boeing 707, returning Moroccan workers home after vacation in France, plunged into mountainside; all 188 aboard killed
> September 10, 1976 — Zagreb, Yugoslavia: midair collision between British Airways Trident and Yugoslav charter DC-9; fatal to all 176 persons aboard
> March 27, 1977 — Santa Cruz de Tenerife, Canary Islands: Pan American and KLM Boeing 747 collided on runway; all 249 on KLM plane and 333 of 394 aboard Pan Am jet killed; total of 582 is highest for any type of aviation disaster
> September 25, 1978 — San Diego, CA: Pacific Southwest plane collided in midair with Cessna; all 135 on airliner, 2 in Cessna, and 7 on ground killed for total of 144
> May 25, 1979 — Chicago: American Airlines DC-10 lost left engine upon take-off and crashed seconds later, killing all 272 persons aboard and three on the ground in worst U.S. air disaster
> June 23, 1985 — Coast of Ireland: Air-India Boeing 747 exploded over Atlantic; all 329 aboard were killed
> August 12, 1985 — Japan: Japan Air Lines Boeing 747 crashed into a mountain, killing 520 of the 524 aboard
> August 16, 1987 — Detroit: Northwest Airlines McDonnell Douglas MD-30 plunged to heavily traveled boulevard, killing 156; girl, 4, only survivor
> December 21, 1988 — Lockerbie, Scotland: a New York-bound Pan Am Boeing 747 exploded in flight from a terrorist bomb and crashed into Scottish village, killing all 259 aboard and 11 on the ground; passengers included 35 Syracuse University students and many U.S. military personnel
> May 11, 1996 — Everglades, FL: ValuJet flight went down in swamp, killing 110; cargo fire caused by oxygen generators missing safety caps
> July 17, 1996 — Coast of Long Island, NY: a TWA Boeing 747-100 bound for Paris from New York exploded over waters of eastern Long Island and crashed into Atlantic Ocean, killing all 230 aboard
> September 2, 1998 — Nova Scotia, Canada: Swissair flight from New York to Geneva crashed off Canadian coast, killing all 229 aboard; 136 Americans were on the McDonnell Douglas MD-11
> October 21, 1999 — Nantucket, MA: EgyptAir Flight 990 went down in Atlantic; all 217 on board killed in apparently controlled rapid descent; deceased crew members are focus of investigation
> February 5, 2000 — Alaska Airline's MD-88 crashed off coast of California; all passengers killed

Although air travel by company officers and managerial team members is commonplace both on commercial airlines and private air transports, prudent safety professionals may want to evaluate the potential risks, however small, and provide appropriate protections.

Dams

If the operation or facility is located downstream from an earthen dam or other dam structure, safety professionals may want to consider assessing the potential risk of an unexpected dam collapse. For example, the disaster at Buffalo Creek in West Virginia was the result of a weakened earthen dam collapse. The resulting rush of water caused substantial damage to the town and resulted in the loss of life as well as property.

Shipwrecks

If the operation involves shipping of products or personnel, safety professionals should consider including an assessment of the risks of such travel. Although travel by sea has become substantially safer in past years, the potential of natural risks, such as hurricanes, and man-made risks, such as terrorism, can increase the risk probability. Some of the more recent disasters at sea include the following:

>   March 9, 1987 — British ferry capsized after leaving Belgian port of Zeebrugge with 500 aboard; 134 drowned; water rushing through open bow is believed to be probable cause
>   December 20, 1987 — over 4000 killed when passenger ferry *Dona Paz* collided with oil tanker *Victor* off Mindoro Island, 110 miles south of Manila
>   April 7, 1990 — *Scandinavian Star*: suspected arson fire aboard Danish-owned North Sea ferry killed at least 110 passengers in Skagerrak Strait off Norway
>   September 28, 1994 — *Estonia*: passenger ferry capsized off coast of Southwest Finland and sank in a stormy Baltic Sea; only about 140 of the estimated 1040 passengers aboard survived
>   February 1999 — *Harta Rimba*: ship sank in the South China Sea, killing about 325 people; the ship had not been licensed for passenger use

Mine explosions

Although the Mine Safety and Health Administration governs underground mining, safety professionals with underground or open pit mining operations should consider including a risk assessment for these operations in the overall assessment.

Chapter two: Natural risks

Railroad accidents

If the transportation of product or personnel is performed via rail, safety professionals should consider inclusion of a risk assessment of this area. Although travel by rail is exceptionally safe, major disasters resulting in the loss of property and life do occur. Some of the most recent accidents involving trains include the following:

> May 3, 1962 — Tokyo: 163 killed and 400 injured when train crashed into wreckage of collision between inbound freight train and outbound commuter train
> November 9, 1963 — Yokohama, Japan: two passenger trains crashed into derailed freight train, killing 162
> July 26, 1964 — Custoias, Portugal: passenger train derailed; 94 dead
> February 4, 1970 — Buenos Aires: 236 killed when express train crashed into standing commuter train
> July 21, 1972 — Seville, Spain: head-on crash of two passenger trains; 76 killed
> October 30, 1972 — Chicago, IL: two Illinois Central commuter trains collided during morning rush hour; 45 dead and over 200 injured
> September 22, 1993 — Mobile, AL: Amtrak's *Sunset Limited*, en route to Miami, jumped rails on weakened bridge that had been damaged by a barge, and plunged in Big Bayou, killing 47 persons
> June 3, 1998 — Eschede, Germany: *Inter City Express* passenger train traveling at 125 mph crashed into support pier of an overpass, killing 98; crash may have been caused by a defective wheel
> October 5, 1998 — London: worst British train crash in quarter century claims up to 40 lives; accident blamed on poorly visible signal that had been missed by drivers eight times in the past four years

The above are several of the potential areas of natural and man-made risks that should be considered in performing the overall risk assessment for the operation or facility. Prudent safety professionals should carefully evaluate all potential lists and appropriately assess the risk for inclusion in their proactive prevention programs as well as within the overall emergency and disaster preparedness planning. No risk is too small to be excluded without at least a preliminary assessment. We should use Murphy's Law as the basis for our assessment — the one potential risk that we exclude for being too remote will be the one risk that will come to fruition.

chapter three

Emerging risks

> "All business proceeds on beliefs, on judgments of probabilities, and not on certainties."
>
> Charles William Eliot
>
> "You never accumulate if you don't speculate."
>
> David Dodge

In recent years, several new and emerging risks have surfaced that require the attention and assessment of the safety professional including the areas of workplace terrorism, workplace violence, bioterrorism, and cyberterrorism. Our changing workplace, with new technologies and global economies, has opened the pandora's box to new risks which were virtually unheard of in past years. Our employees and team members of today are substantially different from their fathers and grandfathers and the expectations of today's workforce is substantially different. Our direct link society, with CNN and instant access internet, has opened the door to the potential of incidents performed to acquire media attention for a cause or fringe position. Safety professionals today must assess these potential risks to their operations and facilities and take appropriate measures within the framework of their proactive and preventative designs to minimize these new and emerging risks.

Terrorism is defined as "the threat or use of violence against the civilian."[1] In recent years, terrorism of various types has become far more prevalent and visible in the U.S. than in years past. As discussed further in this chapter, terrorism can take many forms but the basic premise is to strike terror and instill fear into the hearts and minds of a civilian population. A few of the most recent incidents of terrorism in the U.S. include the following:

> January 24, 1975 — New York City: bomb set off in historical Fraunces Tavern killed 4 and injured more than 50 persons. Puerto Rican nationalist group (FALN) claimed responsibility and police tied 13 other bombings to it.

[1] Infoplease.com dictionary.

February 26, 1993 — New York City: bomb exploded in basement garage of World Trade Center; killed six and injured at least 1,040 others. Six Middle Eastern men were later convicted in this act of vengeance against the Palestinian people. They claimed to be retaliating against U.S. support for the Israeli government.

April 19, 1995 — Oklahoma City: car bomb exploded outside federal office building, collapsing walls and floors; 168 persons were killed, including 19 children and one person who died in rescue effort. Over 220 buildings sustained damage. Timothy McVeigh and Terry Nichols later convicted in the anti-government plot to avenge the Branch Davidian standoff in Waco, TX exactly 2 years earlier.

Organizations or facilities may be at risk of a terroristic attack depending on the nature of the business, the visibility of the organization, and other related factors. Prudent safety professionals should assess the risk probability of their organization and facilities, as well as the operations and facilities located near their operations, for the potential of terroristic activities.

Violence in the workplace

One of the recently emerging risks that should be assessed by safety professionals is violence in the workplace; whether initiated by members of the workforce or not, employees and management team members are at risk. Workplace violence has fast become the leading cause of work related deaths in the U.S. and has opened an expanding area of potential liability against employers who failed to safeguard their workers. According to the statistics from the National Institute of Occupational Safety and Health (NIOSH), over 750 workplace killings a year have been reported in the 1980s.[1] Additionally, according to the National Safe Workplace Institute, there were approximately 110,000 incidents of workplace violence in the U.S. in 1992. A common misconception is that violent incidents are a fairly new phenomenon; however, incidents of workplace violence have been happening for a substantial period of time. The primary reason for the emphasis in this area at this time is because of the increased frequency and severity of the incidents of workplace violence.

According to the U.S. Bureau of Labor Statistics, there were 1063 homicides on the job in 1993 and of these deaths, 59 were killed by co-workers or by disgruntled ex-employees.[2] This report also noted that there were 22,396 violent physical acts which occurred on the job in 1993, and approximately 6% of these incidents were committed by present or former co-workers. In addition to the incidents of workplace violence among and between employees and ex-employees, incidents of other individuals entering

---

[1] Bensimon, Helen Frank "Violence in the Workplace," *Training and Development* at 27 (January 1994).
[2] Census of Fatal Occupational Injuries, Bureau of Labor Statistics, U.S. Department of Labor, August 1994.

Chapter three: Emerging risks

into the workplace, such as disgruntled spouses, have drastically increased. Another area that should be considered within the realm of workplace violence is the sabotage and violence directed at the company by outside organizations. Examples of such incidents are the World Trade Center bombing and the bombing of the Federal Building in Oklahoma City.

Incidents of workplace violence have been highly publicized. The most visible organization with a substantial number of workplace violence incidents is the U.S. Postal Service which recorded some 500 cases of workplace violence toward supervisors in an eighteen month period in 1992 and 1993.[1] Additionally, the U.S. Postal Service also recorded 200 incidents of violence from supervisors toward employees. Below are just a few of the other highly publicized incidents which resulted in injury or death to individuals:

- The shooting spree at the Chuck E. Cheese restaurant in Denver, CO in which a kitchen worker killed four employees and wounded a fifth.
- The ex-employee of the Fireman's Fund Insurance Company who killed three individuals, wounded two others, and killed himself in Tampa, FL.
- The 1986 Edmond, OK shooting where a letter carrier killed fourteen and wounded six others.
- The disgruntled postal worker in Dearborn, MI who shot another employee in May 1993.
- The former postal worker who killed four employees and injured another in the Montclair, NJ post office.

So what exactly is workplace violence? Generally, workplace violence is defined as "physical assaults, threatening behavior, or verbal abuse occurring in the work setting".[2] Although incidents of threatening behavior such as bomb threats or threats of revenge are not statistically available, there is a substantial likelihood that these types of incidents are also on the increase.

Many companies and organizations in the U.S. have taken steps to safeguard their employees in the workplace through a myriad of security measures, policy changes, and other methods. The potential legal liabilities in this particular area have drastically increased for employers. In most circumstances, the employer would be responsible for any costs incurred by the employee through the individual state workers' compensation system. Now, however, new and novel theories such as negligent retention, hiring, and training, as well as the potential of governmental monetary fines leveled by OSHA, have also merged to increase the potential risk.

Most experts concede that there are no magic answers when it comes to addressing problems in the area of work related violence. Given the fact that the potential of violence exists on a daily basis and that violent methods can

---

[1] Kurlan, Warren M., "Workplace Violence," *Risk Management*, at 76 (June 1993).
[2] Physical assault, from our research, has run the gamut from an employee shoving or punching an employee to the use of a weapon or explosive to kill the individual.

be precipitated can come from a wide variety of areas, the intangibles lend themselves to the fact that workplace violence is a very complicated issue. Incidents of workplace violence have been occurring since the Industrial Revolution in the U.S. The number of incidents has increased substantially as well as the severity of these types of workplace incidents. This may correlate to a variety of reasons including, but not limited to, the increased violence in our society, the availability of weapons, the down-sizing of the workplace, the management style, and many other reasons. Additionally, when you include the different types of workplaces in America as well as the variety of management approaches, there is no one simple answer to this multifaceted question.

The Occupational Safety and Health Administration has provided guidelines for specific industries such as the retail industry and health care operations.[1] Many employers have taken proactive steps to develop a general strategy in order to protect their employees and thus reduce the potential legal risks as well as providing ancillary efficacy benefits to employees and management. In addition to the proactive strategy, many employers have developed a reactive plan and implemented stringent employee screening and monitoring processes to identify and address potential incidents of workplace violence in order to minimize potential risks.

In most circumstances, employers are better able to combat the potential risk of workplace violence when the threat is initiated by an employee rather than an ex-employee or outside individual. Researchers have provided a general profile of individuals with a propensity toward workplace violence which include employees with depression, suicidal threats, poor health, and other traits. Incidents precipitated by ex-employees, spouses of employees, and individuals outside the organization are substantially harder for the employer to address.

As safety professionals attempt to address the potential risks of workplace violence and the correlating legal risks and costs, the employers must be very cautious not to trample upon the individual's rights and freedoms. As employers develop and implement more stringent activities and programs to curtail or minimize the potential risks of workplace violence, they must be extremely cautious to not create additional legal risks through their actions. Privacy laws, acquisition of information laws, and discrimination laws provide avenues of potential redress in this area.

Workplace violence is a risk which should be appropriately addressed by safety professionals as part of their proactive emergency and disaster preparedness efforts. Incidents of workplace violence possess virtually all of the components of an organizational planning program. The Occupational Safety and Health Administration has provided guidance as to the appropriate measures to be addressed if workplace violence is a viable risk in your organization (see Appendix A).

[1] Workplace Violence: OSHA says guidelines will target the retail and health care sectors, 1995 DLR 16 (BNA, January 23, 1995).

Cyberterrorism

In today's information age, knowledge is power. Virtually every organization and company utilizes computers to increase productivity, to communicate between units, and to access information. Computers have made our work easier but also have created a new way for individuals with a computer to potentially disrupt and cause havoc within an organization. Additionally, with the explosion of the "dot com age," many organizations and individuals are now providing financial and personal information through the internet, which carries with it a potential risk that was not present even a short 20 years ago.

One of the newest risks that safety professionals should assess is the potential of cyberterrorism. As noted in a recent CSIS report,

> In today's electronic environment, many haters can become a Saddam Hussein and take on the world's most technologically vulnerable nation. America's adversaries know that the country's real assets are in electronic storage. Virtual corporations, electronic transactions, and even economies without inventories — based on just-in-time deliveries — will make attacks on data just as destructive as attacks on physical inventories. Bytes, not bullets, are the new ammo. Or, most dramatically, a combination of bytes, bullets, and bombs.[1]

Although the issue of cyberterrorism may be foreign to many safety professionals, most organizations possess individuals or teams of information specialists or computer specialists who are very familiar with the operation of the computer systems. It is important for the safety professional to become generally educated in the operations of these systems and apply the basic principles of safety to the these new information systems in order to properly assess the risk factors involved. As a safety professional you may possess a limited knowledge of your program's information system. However, utilizing the basic safety and risk assessment principles, the safety professional can ask the appropriate questions of the computer experts to acquire the information necessary to assess the current situation. For example, you would not want employees to be able to peruse your personnel files, so procedures and barriers are placed between the employees and this information — the same with the computer information. Are appropriate fire walls in place to prohibit individuals from acquiring sensitive information from your information system? If an employee with an airborne communicable disease wanted to slow your operation by making employees sick, he or she would spread this virus through the air system. If an employee wanted to cause havoc with your information system, he or she could plant a virus which could slow or destroy your data base.

[1] CSIS Press Release, December 15, 1998, p. 1.

Cyberterrorism by disgruntled employees or ex-employees, terroristic organizations, your local hacker, or anyone with a computer and a grudge can be a substantial risk for an organization or company. Although not within the usual purview of the safety professional, this emerging risk is an area to consider for assessment if your organization is a target for cyberterrorism. The aftermath of a cyber attack can detrimentally affect the foundational elements in which the organization functions and the financial health of your organization.

Bioterrorism

One of the new potential risks that should be considered within your risk assessment is bioterrorism. Depending on the type of operation, facility location, and other related factors, facilities and organizations may be in direct or peripheral risk of harm. Bioterrorism is the risk of the use of chemicals, biological agents, radiological, or even nuclear materials to inflict harm on individuals or organizations. As we have seen from the various incidents throughout the world, access to chemical or biological agents is substantially easier today than in the past, and the technology and information to transform these agents into weapons is available at the nearest computer terminal.

Although the risk of an actual bioterrorism event at most facilities is relatively small, the risk of a hoax or threat of a bioterroristic event can cause substantial disruption as well as related potential risk to employees and the organization. Not unlike a bomb threat, employees, ex-employees or virtually anyone with a telephone or access to e-mail can trigger a reaction by the organization causing substantial losses. Furthermore, in the event of an actual bioterroristic event, virtually no private sector organization is prepared to address the needs of the situation without substantial preparation.

The potential of internal sabotage is an area of risk among companies utilizing chemicals, gasses, and other volatile substances in the product development process. Safety professionals may want to evaluate the potential risk of internal terroristic risks or internal sabotage which could create substantial risks. For example, XYZ company makes Product A utilizing a flammable chemical. Employee Z just received disciplinary action for a workplace violation. Upon the employee's return to the workplace, he or she simply leaves the valve to the flammable chemical open at the end of the shift to permit the chemical to overflow and be exposed to an ignition source thus starting a fire in the facility. Is this an accident or an incident of deliberate sabotage?

In summation, safety professionals may wish to assess the potential risk of these new and emerging areas within the framework of their overall analysis. In most circumstances, the risk of bioterrorism is substantially low. Cyberterrorism is very real, as seen from the recent attacks reported on CNN and NBC; however, additional expertise may be necessary to make an appropriate assessment. Workplace violence is a new arena which possesses substantial information and should be an area of assessment for any workplace with employees. Today's safety professional faces potential risks virtually nonexistent in the not-so-distant past. New potential risks that we have not even fathomed will be emerging in the future. Prudent safety professionals should be vigilant and ensure all potential risks are properly assessed, analyzed, and addressed.

chapter four

Governmental regulations

"Law cannot persuade where it cannot punish."

Thomas Fuller

"That law may be set down as good which is certain in meaning, just in precept, convenient in execution, agreeable to the form of government, and productive of virtue in those that live under it."

Francis Bacon

On a federal level there are three primary governmental agencies tasked with regulatory responsibilities in the emergency and disaster area. The primary agency responsible for regulations and response for public emergencies is the Federal Emergency Management Agency, or commonly referred to as FEMA. This agency is responsible for a wide range of emergency related responsibilities ranging from flood insurance to response following a large scale disaster situation.

In the private sector, the primary agency which promulgates regulatory requirements is the Occupational Safety and Health Administration, or OSHA. Additionally, there are several overlapping regulations promulgated by the Environmental Protection Agency or EPA, which also come into play with an emergency and disaster preparedness plan. Safety professionals must ascertain which regulations are applicable to the circumstances and ensure compliance with the appropriate regulation. To assist in this assessment, OSHA provides several publications that can be acquired on their website at **www.osha.gov** or from their publication operations department. EPA information can be acquired at **www.epa.gov**.

In most circumstances, safety professionals find that an assessment of the operations or facility can be conducted initially and the appropriate regulation is located after the potential hazards or circumstances of the operations or facility are assessed. The primary areas of concern in conducting the initial facility or operation assessment should include, but are not limited to, the following areas:

- Type of operations
- Chemicals/operations
- Evacuation routes
- Emergency lighting
- Emergency response
- Fire extinguishers
- Communications
- Designated areas and accounting for employees
- Triage areas
- Transport
- Public sector response
- Highways and clearance for emergency vehicles
- Media control
- Information dissemination
- Medical equipment and personnel
- Control valves or control equipment
- Internal v. external evacuations
- Selection of personnel
- Training of personnel
- Specialized equipment
- Specialized procedures

Upon completion of the entire operation or facility assessment, the emergency and disaster preparedness plan should identify the various elements to be addressed in the development of the plan. Safety professionals can then conduct a search of the literature and regulations to identify all applicable regulations and research that address each of the identified elements to be included in the plan. Some of the common regulations to be assessed include the following:

EPA (EPCRA)

The Emergency Planning and Community Right-to-Know Act (EPCRA), enacted in 1986, has two major purposes: (1) to increase public knowledge of and access to information on the presence of toxic chemicals in communities, releases of toxic chemicals into the environment, and waste management activities involving toxic chemicals; and (2) to encourage and support planning for responding to environmental emergencies.

OSHA standards

1. Hazardous waste operations and emergency response   1910.120
2. Hazardous waste operations and emergency response   1926.65
3. Training curriculum guidelines — non mandatory   1926.62

Chapter four: Governmental regulations

4.	Training curriculum guidelines — non mandatory	1910.120
5.	Fire protection	1910 Subpart L App. A
6.	Means of egress	1910 Subpart E
7.	Flammable and combustible liquids	1910.106
8.	Existing installations — mandatory	1910.66 App. D
9.	Employee emergency plans and fire prevention plans	1910.38
10.	Emergency action plans	1918.100
11.	Respiratory protection	1910.134
12.	Emergency action plan	1917.30
13.	Process safety management of highly hazardous chemicals	1910.119
14.	Model emergency temporary standard	1990.152
15.	Employee emergency action plans	1926.35
16.	Flammable and combustible liquids	1926.152
17.	Employee alarm system	1910.165
18.	Fire brigades	1910.156
19.	Emergency standards	1911.12
20.	Portable fire extinguishers	1910.157
21.	Hazardous atmospheres and substances	1917.23
22.	Means of egress	1910.37
23.	Emergency disclosures	71.11
24.	Access to employee exposure and medical records	1910.1020
25.	Signs and markings	1917.128
26.	Fire detection systems	1910.164
27.	Means of egress	1926.34
28.	Changes to approved plans	1952.297
29.	Level of federal enforcement	1952.272
30.	Level of federal enforcement	1952.265
31.	Description of the plan as initially approved	1952.290
32.	Description of the plan	1952.270
33.	Description of the plan as initially approved	1952.260
34.	Description of the plan as initially approved	1952.250
35.	Completion of developmental steps and certification	1952.232
36.	Description of the plan as initially approved	1952.240
37.	Completion of developmental steps and certification	1952.212
38.	Developmental schedule	1952.211
39.	Description of the plan as initially approved	1952.210
40.	Description of the plan as initially approved	1952.200
41.	Level of federal enforcement	1952.172
42.	Description of the plan	1952.170
43.	Description of the plan as initially approved	1952.150
44.	Description of the plan as initially approved	1952.160
45.	Power transmission and distribution (emergency lighting)	1926.950

As can be seen, there are a number of potentially applicable OSHA standards and EPA regulations that may be required within your overall emergency and disaster preparedness program. Safety professionals are urged to be diligent in identifying all applicable current regulations as well as proposed regulations in the development of the plan. Additionally, safety professionals should ensure that each and every element within each regulation is appropriately addressed within the emergency and disaster preparedness plan. Compliance with all OSHA, EPA, and other applicable regulations and standards is essential in order to avoid potential penalties and litigation as well as to ensure a complete and thorough emergency and disaster preparedness plan.

chapter five

Structural preparedness

> "Action without study is fatal. Study without action is futile."
>
> Mary Beard

> "Unless there be correct thought, there cannot be any action, and when there is correct thought, right action will follow."
>
> Henry George

Once the risks and probabilities of disasters have been identified, physical facilities must be examined in detail and appropriate countermeasures implemented to prevent and minimize the impact of unwanted emergencies. The age and condition of the physical plant will determine the size and scope of this task. A new complex with all new buildings, constructed in compliance with the latest building code would present less of a challenge than an old facility constructed in many phases. The level of commitment, or lack of commitment, to maintaining existing facilities will become apparent as the inspectors proceed. In older buildings that have housed several tenants and many modifications, basic structural integrity may have been compromised. On one inspection of an older facility, the interior of a large warehouse had been renovated into two floors of office space for design and production personnel. A major structural component supporting the roof of the building had been severed to add a door, enabling the space on the other side to be accessed. The bottom chord of the roof truss had been removed without adding reinforcement. No engineering studies were conducted to determine the impact prior to construction. The fire protection system of the building had not been modified to add additional heads in this new building built inside the original one. The company maintenance manager assured me that design safety factors and the adjacent trusses would support the roof. Under normal circumstances, time was proving his point; however, severe snow load, a blocked roof drain, or strong winds could have easily collapsed the roof with high dollar losses and the potential for death and injuries. A fire separation wall constructed at considerable cost to obtain a two or four-hour rating is easily compromised when an opening is made to allow people or vehicles access. The decision makers may not have even realized it was a rated wall. Other examples of building modifications that have the potential to increase fire spread and endanger occupants include:

- Increased travel distance to exits beyond legal limits imposed by codes
- Improper storage of materials including flammables, combustibles, and toxic materials
- Overloaded electrical circuits and/or lack of emergency lighting in renovated areas
- Absence of firewalls and/or fire partitions
- Failure to extend firewalls and/or fire partition walls beyond the underside of a ceiling
- Penetration of fire separation walls by utilities (telephone, cable, data-transmission lines)
- Fire doors compromised by physical damage or propped open with wedges

Although good housekeeping does not ensure a building's fire safety, this author has never seen a facility with poor housekeeping that was fire safe. If you find a facility with large quantities of combustibles accumulating and in desperate need of cleaning, closer examination will probably reveal larger maintenance and fire protection deficiencies. This first step in preparing the physical facilities for a disaster is to determine their structural integrity on a daily basis, without the addition of an emergency. Technical means alone cannot accomplish this task for you; however, you should use all means available to assist your survey. The availability of thermal imaging equipment is an example of a tool that can assist the inspector in finding air, steam, and water leaks as well as heat or cooling losses. Finding leaks early while they are small saves energy and reveals weak spots in the integrity of these systems that could fail under emergency conditions.

Another area frequently overlooked is fire protection system integrity. Have the fire detection and suppression systems been inspected and maintained according to the applicable National Fire Protection Association Standards? If the occupancy use of a structure changes, the fire protection system may not be able to contain or extinguish a fire. An automatic sprinkler system designed to handle the fire load of a light hazard office occupancy probably could not handle the fire if the space was converted to records storage (large quantities of paper) or flammable liquids storage. This may seem obvious to modern safety professionals, but at some point in the past, the decision to convert the space may have been made without the advice and counsel of safety and/or fire protection personnel. Another classic example often used is a hypothetical XYZ tricycle manufacturing plant. The tricycle

Chapter five: Structural preparedness

designed and manufactured during the 1940s and '50s was mostly steel with padded seats and rubber tires. If the XYZ tricycle manufacturing plant was fortunate enough to have installed an automatic sprinkler system then, that same system would be ineffective in protecting a tricycle built by today's standards. The tricycle manufactured today is almost entirely plastic and is even stored in corrugated cardboard boxes. The fire load imposed by the plastic and cardboard would probably be too much for the XYZ company's sprinkler system designed for the manufacturing of the '40s or '50s. How old is the sprinkler system in the building(s) you protect? Has the system been upgraded to reflect increased use of plastics and/or flammable/combustible liquids? A thorough analysis of the condition, size of the sprinkler piping, heads, and the water supply by a sprinkler design engineer will yield a plan to restore the system's effectiveness. In some cases, if the water supply is strong enough, the heads are located properly, and the piping is of sufficient size and condition, the system can be upgraded by simply replacing the sprinkler heads with larger diameter sprinkler heads.

Before leaving the subject of automatic fire protection systems, earthquake protection, control valve supervision, and protection of systems during cold weather should all be evaluated. The National Fire Protection Association Pamphlet 13 Standard for the Installation of Sprinkler Systems includes a seismic map in the appendix showing the potential for earthquake damage and the need for protection of the piping from earthquake damage.

Some older systems may not be designed to withstand an earthquake and could leave the facility vulnerable at a time when post quake fires are common. All fire control valves must be of the indicating type and must be locked in the open position and/or electronically supervised. Most insurance companies have programs designed to minimize the time fire control valves are closed and ensure systems are promptly restored to operation after any maintenance or sprinkler operation. The level and temperature of water in storage tanks and the status of fire pumps must also be monitored to ensure their operation if needed. Buildings with sprinkler systems have burned to the ground simply because a valve was accidentally or intentionally closed.

Cold weather poses challenges to improperly designed or maintained sprinkler systems. Wet pipe systems should not be installed on loading docks, in freezers, or anywhere temperatures may drop below 40°F. Each fall, all low point drains on dry pipe systems must be checked for accumulation of water. Antifreeze based sprinkler systems must also be checked each fall before freezing weather to ensure the solution will protect against the lowest expected temperature.

Structural security from flash floods, mudslides, and high water of any type should be assessed and appropriate intervention methods developed long before the need for those plans arises. What utilities might be compromised? Are there any critical components such as pumps, transformers, raw materials, or finished goods that should be relocated permanently to prevent losses, or do you have sufficient resources to move and/or protect those items should the situation dictate it? One industrial plant located on a small island lost its only evacuation route when a toxic chemical release enveloped the only bridge linking the island with the mainland. Contingency plans now include boats to egress people and bring in emergency response personnel and assistance.

Building security must also be evaluated and the risk and probability of disasters from attacks by external or internal sources must be assessed. This should include threats to any and all of the company resources including:

- Property
- Employees
- Equipment
- Inventory
- Data
- Trade secrets
- Manufacturing methods and procedures
- Company image, reputation, and consumer confidence

The loss of consumer confidence after the Tylenol product container tampering served as a wake up call for many companies. Most companies responded quickly, and today most products have packaging seals to warn the consumer of product tampering. At times prudent security measures seem to run counter to fire and emergency evacuation procedures. Computer integrated access control systems are being viewed as a valuable tool in this battle. Photograph identification badges with cameras at all entrance/exit points limit unauthorized personnel from company property. Access control locks operated by the same badges limit access to those areas each employee should enter. The computer integration of the locks also maintains a record, including dates and times, of all employees entering and leaving various areas. This same type of system is very helpful for accountability of employees when an evacuation is necessary. The computer can instantly generate a list of all employees who have not reported to an evacuation checkpoint.

Workplace violence remains a threat to the personnel and success of businesses worldwide. A domestic dispute in a factory in Kentucky ended with a high speed vehicle crash through the main employee entrance. The car crashed completely through a cafeteria that only minutes earlier had been filled with employees. Every facility must have a workplace violence plan. Some companies choose to make it part of the emergency action plan and others choose to make it a separate plan. In either case, it should focus on preventing and settling disputes while they are manageable and protecting the employees in the event a violent act does occur. Physical barriers that prevent vehicles from being driven into selected areas can be incorporated into the design of parking areas. Advances in closed circuit video camera technology continue to increase this tool as an effective means to deter unwanted persons from company property.

Chapter five: Structural preparedness 37

The role that safety and/or security forces must play in emergency response to fires, medical emergencies, hazardous material incidents, confined space rescue, and bomb threats should be based upon careful analysis of:

1. The proximity of well-trained, well-equipped external assistance
2. The level of automatic protection available
3. The potential for loss if external assistance is insufficient
4. Management commitment to training and equipping an internal team

Well intentioned people frequently say things like "I would never have fire brigades. They are too expensive." If your facility is located in a part of rural America with little or no organized emergency response near the plant, how will you prevent the plant from burning down if you don't have a functional fire brigade? Another common response is "We don't have a fire brigade. We have a hazardous response team." The bottom line is **the mission statement for your team (whatever you choose to call them) should describe in detail the duties they should and should not perform**. If their duties include the use of any fire protection equipment, then they must receive the training and protective equipment consistent with that level of response regardless of the name given the team. The nature of emergency response lends itself to having a multi-purpose team. A medical screening team ensures the employees are physically fit to wear respirators, have incident command and respirator training, and understand the chemistry of fire. Investing company time and money in training a single team to deal with all emergencies makes good sense. Most plants also have civic minded employees who already serve on local fire departments and/or rescue squads. These people have experience, may be willing to serve on your team, and may already have some of the required training. Often these people have the unique skills and leadership qualities to organize, train, and serve as officers of industrial emergency response teams. Any additional training you provide them benefits your facility and the community at large. For those who choose to evacuate only and wait for the insurance money to rebuild the facility, ask yourself this question: Can other divisions of the company satisfy our customers' needs until the plant is rebuilt or will the customers use another supplier until and when our facility is rebuilt?

The level of emergency response could range from a small group of maintenance personnel trained to use fire extinguishers and provide information to assist the local fire department to a full interior structural attack brigade trained and equipped equal to their municipal counterparts. The decision should be based upon an analysis of the risks and potential consequences.

chapter six

Coordinating with local assets

> "No matter how much work a man can do, no matter how engaging his personality may be, he will not advance in business if he cannot work through others."
>
> John Craig

> "Coming together is a beginning, keeping together is progress; working together is success."
>
> Henry Ford

The ability to communicate, coordinate, and work effectively as a team can be a major factor in the success of your emergency plan. In a large-scale disaster, more and more resources will be needed. The ability of those responding agencies to understand the scope and size of the problem and their ability to communicate among themselves will also influence the success of the operation. The more familiar they are with your facility and the more they have worked together, the more effectively they will be the day you need them most. Many of the agencies you will work with exist to serve the community. Law enforcement, firefighters, paramedics, and other public safety personnel want to protect the community they serve and to protect the property, jobs, and taxes the company generates. Their level of preparedness to help you will vary from community to community. It is your job to assess their capabilities and provide them with the training and special equipment they need to succeed in protecting your assets. It may take some selling on your part to convince them your facility has unique challenges and merits special attention on their part. Meet with the local fire chief and discuss your desire to involve his department in the planning process. The same invitation should be extended to the police chief and the directors of the local emergency medical service agency. If your facility or the surrounding area must be evacuated quickly, you may also want to involve a local transportation agency or the school board to provide bus transportation. As you prepare the emergency response plan, you may discover the need for specialty equipment such as sand or sand bags, dump trucks, specialty law enforcement teams such as SWAT, or bomb technician units. Some of the agencies you contact may feel they are already adequately trained and equipped to handle emergencies at your facility. It is your job to show them their need to walk your facility and study the activities and processes and then decide if they are properly trained and equipped to respond effectively. Because many fire and EMS personnel work a three platoon system you may have to conduct several tours to enable all the agency personnel to participate. You may be fortunate enough to work with responding agencies who are well read and may know as much or more about the historical nature of emergencies in facilities like yours than you. You could also encounter responding agency personnel with good intentions but little knowledge of the potential for death, injuries, environmental impact, or economic losses at your plant. You must play with the hand you have been dealt. Seek out those individuals who have the best grasp of the situation and enlist their help in educating the others. Do not be afraid to describe the negative impacts on their agencies and the community if emergency response is not effective.

The LEPC (Local Emergency Planning Committee) will be another asset you should draw upon in the creation of your emergency response plan, and the SERC (State Emergency Response Commission) will want to evaluate your plan. Depending upon the nature of your business and the quantity of various regulated hazardous materials at your facility, you may be regulated by the Process Safety Management Standard.

Once you have enlisted the help of a team with representatives from these responding agencies, begin to work through all of the potential emergencies your risk analysis identified. Conduct a resource list based upon the answer to a question repeated over and over: If this happens, then what will I need? As you identify each of these resources needed, have someone contact the appropriate business or agency and negotiate until the resource is available. Among those on this resource contact list you may find:

- Emergency sheltering facilities
  - Hotels
  - Motels
  - School gymnasiums
  - National Guard Armories
  - Public arenas
- Community outreach groups
  - American Red Cross
  - Salvation Army
- Heavy construction equipment companies
  - Trucks
  - Cranes

Chapter six: Coordinating with local assets 41

- Contractors
  - Building supply companies
  - Sand
  - Fly ash
- Fire protection foam suppliers
- Equipment rental companies
  - Generators
  - Pumps
  - Heaters
- Public works department
- Utility companies
- Public health department
- Hospitals
- Medical helicopters
- Coroner
- Bomb technician team
- SWAT team
- Coast Guard
- National Spill Response Team
- National Weather Service
- CHEM TREC
- ATF (Alcohol Tobacco and Firearms)
- FBI (Federal Bureau of Investigation)
- FAA (Federal Aviation Authority)
- Critical incident stress debriefing counselors
- Insurance carriers

The size, scope, and nature of your facility and its proximity to major population centers, lakes, rivers, railroads, airports, and airways will impact the nature of your resource list. Hopefully you will never need any of these resources, but plan as though you will need them.

Once the resource list is constructed turn your attention back to those initial response agencies that you know will be responding to emergencies in your facility. Once you have identified all the agencies on your team, get to know them. This emergency response team in many ways is like any organized sports team. Some players will be better than others and some bring unique and special equipment, talent, and capabilities to the team. Your job as their coach is to save your facility by helping them prepare and train for the emergency you hope they never have to tackle. Another challenge you face is the fact that your facility is just one of many they protect. These response agencies have a community to protect and your personnel have products to produce or services to provide, so you can't spend all your time preparing for emergencies. As a team you must decide the level of response you will have to various emergencies. Concerning hazardous material emergencies, you will have to train your team to at least the First Responder Awareness Level. If the agencies that protect your facility do not already have an organized Hazardous Material Response Team, you may elect to train and equip selected groups to more involved levels such as:

- First Responder Operations Level
- Hazardous Material Technician
- Hazardous Material Specialist

The awareness and operations levels are considered defensive and take no action to clean up spills. They concentrate on preventing the spread and protecting unaffected areas. Teams working at the technician and specialist levels have protective equipment and training to enable them to successfully enter the spill areas (hot zone) and safely work to clean up the particular release. The level of training and equipment required increases with each additional level. Technicians operating at the operations level or above must be directed by a commander who has successfully completed Hazardous Material Scene Manager Training. Individuals trained at each of these levels require refresher training to maintain their skills and competencies. Again, you may be fortunate enough to have a well-equipped and well-trained HazMat team protecting your facility. These individuals will welcome the opportunity to learn as much as possible about your facility because they know they will be called upon to mitigate any release. If you do not have a team in place, you may be able to provide financial support to train and equip the local fire department. Again, this is an opportunity to be a good corporate neighbor since the training and equipment you supply will be available for the entire community.

Time is a resource that is limited. Make the most of meetings, training, and planning sessions by developing agendas and calling to remind all agencies involved to ensure that activities start and end on time. Because these are emergency response agencies, recognize that some response companies may not show up or may be called away because of another emergency. If you rely heavily on volunteer responders, most of your planning and training may have to occur in the evenings when most of the members are not working.

The success of your emergency planning and the potential future of your facility will depend to a great extent upon your ability to motivate and encourage this energetic group of brave public servants. Your coaching efforts must continue through all aspects of researching, writing, implementing, training, and rehearsing for the emergency you hope never happens.

chapter seven

Pre-planning for a disaster

"The man who is prepared has half his battle fought."

Miguel De Cervantes

"I thatched my roof when the sun was shining, and now I am not afraid of the storm."

George F. Stivers

Internal actions

There are many actions your own employees can take prior to an emergency that will positively impact both their performance and outside agency performance when and if an emergency should arise. This section focuses on those basic pre-planning activities that are not necessarily site specific.

A chain of command chart with roles and responsibilities for each critical position on the chain should be developed by position title and should be listed three positions deep in all critical locations. If the maintenance superintendent is ill or out of town, who should be contacted? Do not place names on the chart because people retire, are promoted, or transfer. Someone must fill the position and that is why we chart position titles. If the organizational structure is modified, the chain of command must be examined and modified as necessary. This chart could be incorporated into the written plan and utilized with an employee emergency contact list to call back critical employees in the event of an emergency.

An employee emergency contact list should be developed alphabetically and also by position title and area of the plant. It may be essential to contact an off duty employee for critical information, or you may have to notify a family member if an employee is injured. A wallet sized list of essential and critical personnel contact information should be developed and disseminated to supervisory personnel.

Teams of employees should be assigned to conduct "what if" studies concerning any operations which have the capability of disrupting production if they fail to perform normally. If too much catalyst is added to the batch process what will happen? If a pump or sensor fails, what will happen? If a storage vessel is overfilled or ruptures, what will happen? For every possible negative event, the impact of that event and the probability of it occurring should be identified. In a baseball game, every defensive player should know before the ball is thrown what they will do if the batter hits the ball to them. This is the purpose of pre-planning. If something goes wrong with this piece of the process due to equipment failure or human error, what can we do? What should we do? The answer may be to evacuate the entire facility to minimize injuries and death, or the answer may be as simple as diverting the product to another container or containment area.

Who has the authority and responsibility to stop a production line, sound a fire alarm, order an evacuation, or contact outside assistance? These questions must be answered prior to the crisis or valuable time will be wasted. In a disaster, seconds can mean the difference between life and death or between a small loss or a major disaster. What role will your safety/security forces play in various types of emergencies? Will they play an active role in mitigating the emergency or will they support external agencies? Will they simply direct, control, and record the movement of emergency personnel? Depending upon the nature of your facility, the level of outside assistance readily available, and their capability, all of these are reasonable choices.

Maps showing various items should be developed in the pre-planning process and be included in the emergency plan. Maps detailing evacuation routes should be posted for emergencies that occur in the facility and should lead employees to safe areas outside. Accountablity stations should be staffed at these sites to determine which employees may still be in the facility. Employees should not be permitted to leave the property until the supervisor has released them. Employees should not be permitted to re-enter the facility unless directed to do so by the incident commander. In the event of a tornado or hurricane, traditional evacuation routes are ineffective and the employees may have to shelter in place. The routes to these shelters on the interior of structures should also be posted. Although evacuation during an earthquake is not advised, employees in many parts of the country must plan for earthquakes too. In general, if employees are outside, they should move away from buildings and items that may fall on them. If indoors, the employees should be directed to take cover under sturdy furniture and along interior walls and corners. After the earthquake, proceed to outside evacuation areas and their accountability stations.

Because many emergencies require large quantities of water as an extinguishing agent, the plan should include a site map of all potable and non-potable water sources, piping, control valves, sprinkler risers, fire hydrants, fire pumps, and switch gear. A separate map should detail all utilities, transformers, and shut-off locations. Wherever the command station is established, security and maintenance personnel should be assigned to assist the emergency scene commander if questions about the facility arise. Additional maps from the engineering and maintenance department should be readily available for the incident commander. These include but are not limited to individual building floor plans, storm drain lines, sewer lines, heating lines, cooling lines, process water lines, and hazardous materials storage areas.

In certain operations, the shut down of equipment may require some crucial employees to remain at their stations as long as safely possible. This will minimize the hazard to those exiting and prevent an even larger catastrophe. These situations must be communicated in the plan to prevent responding agencies from insisting on total evacuation. The response personnel may also be able to assist the employees remaining in their evacuation stations when practical or if conditions become too dangerous for them to remain.

Evacuation alarm information must be clearly communicated in the plan and a suitable system for alerting the employees must be present. In high noise areas, vibrating pagers, high intensity strobe lights, or other means must be utilized to alert occupants of an emergency. The Americans with Disabilities Act has enabled people with hearing, vision, mobility, and other challenges to enjoy working in areas that were considered off-limits a few decades ago. In an emergency, these individuals may need special assistance to ensure their safety. Your plan must provide for their safety.

Standard procedures should be developed to protect valuable computers and the data stored on them in the event of an emergency. Instantaneous standby power capable of operating long enough to back up critical files, shut down essential equipment, and transfer operation to unaffected areas or plants should be in place and tested regularly. The loss of accounts receivable information, inventory, and purchase or delivery orders could result in economic disaster. Internal response to any disaster or emergency should include automatic means of protecting this equipment and data.

A significant portion of any new maintenance person's training should include familiarity with all building systems including fire protection, utilities, electrical distribution, and emergency plan procedures. These individuals should also participate in training and testing of the emergency plan.

Communication plans should be developed instructing employees and their families where to tune their radios and/or televisions to learn about plant closings, re-openings, information about injured employees, etc. This will minimize switchboard traffic for more urgent communication. Some private telephone numbers should be unpublished and retained for essential communication during emergencies.

Your plan should include dispatching personnel to control panels, fire pumps, and control valves to prevent failure or premature closing of valves before the incident commander is certain the fire is out. Just because a light on a panel indicates the pump is running or a valve is open, it doesn't mean the pump is pumping water or the valve is completely open. Always guard the control valves, pump, etc. until ordered to close it by command. The fire literature is filled with accounts of buildings burning down because someone afraid of water damage closed the sprinkler valves. Many people who set fires realize that sprinklers are their enemy and intentionally disable them.

This chapter has primarily addressed responses to natural disasters or emergencies occurring within the facility. Do not overlook the possibility of an emergency off-site impacting your facility. Proximity to highways, rivers, airports, airways, or corporate neighbors could bring a disaster to your doorstep. If your facility is located on the approach path to a busy airport or has a plant with a large quantity of hazardous materials nearby, your facility must incorporate these potential problems into your emergency plan.

External actions

Many actions that occur outside your facility can impact the success of the emergency response. Many of these actions will be outside your control when the emergency happens, but careful planning on your part during the design and implementation of the plan will enable these key actions to be accomplished. Depending upon the size of your facility and the nature of your business, you may have to dispatch security officers to meet responding agencies and escort them to the exact location. This is a good idea regardless of the size of the facility. The responding agency may be staffed with individuals not as familiar with the facility as those normally assigned to the station that protects your property. Other simultaneous municipality emergencies, vacations, illness, and other reasons may result in the least trained people in the department arriving at your gate. In that case, the availability of a knowledgeable security/safety person would be a welcome sight to the arriving firefighters. Hopefully their previous training and the documentation in the emergency plan will provide the necessary information to solve the problem.

Your previous training and the emergency plan will dictate the level of involvement your staff will have with the responding agencies. At the very least, you or your plant emergency coordinator should have a small command staff staged and ready to respond or answer any questions the incident commander may have. They should be located some physical distance from the command post, but accessible by radio.

The television and radio media personnel will report the story of the emergency occurring at your facility. Through the effective use of a public information officer, accurate information can be supplied to the media preventing them from misunderstanding and misrepresenting the size, scope, and magnitude of the event. If you fail to provide accurate information or allow every employee to speak with the media, the story they report may be inaccurate and misleading, and could result in substantial unnecessary negative public opinion. This could even impact stockholders' confidence and impact the market value of the company. If the responding emergency personnel already have individuals trained as public information officers, you may be able to communicate effectively through them.

Large scale emergencies require large numbers of people, vehicles, and support equipment. All of these groups must be staged at a location off-site and released by a staging officer upon the request of the incident commander. It is imperative that a suitable staging location be established quickly and a competent staging officer maintain control. Law enforcement personnel must seal off access to the facility and allow only requested vehicles inside the perimeter. Your emergency plan must incorporate authority for towing vehicles that are blocking emergency vehicles access. At the April 20, 1999, high school shooting in Littleton, CO, several locked, unattended police vehicles had to be towed from the scene to permit ambulance and rescue vehicle access.

If your facility has the potential for mass casualties, you may need a separate staging area for ambulances and air transport. The EMS staging officer will work closely with the triage officer to make sure that transport units arrive as needed and that the most seriously injured victims are treated and transported first. Your plan should examine possible locations for staging apparatus, patient triage, potential landing sites for medical helicopters, and a temporary morgue. Mass casualties will also require large numbers of support people to deal with the questions from family and loved ones inquiring about those potentially killed or injured. Television or radio announcements that direct the individuals away from the scene to a location where accurate information and counseling can be provided will ease congestion at the scene.

Your plan should include the mechanisms that will be used if populated areas near the plant must be evacuated. This could be the most important step taken to reduce death and injuries. Environmental disasters may be minimized if contingency plans are developed for dealing with the possible unexpected rupture of a container or release into the land, water, or air. Technology is helping in this area too. Computer software such as CAMEO that predicts hazardous material vapor cloud size, direction, and speed is already available. Soon the incident commander's laptop PC will communicate via interactive cellular phone networks to their sector commanders in real time using small personal data access (PDA) terminals. Interactive checklists will automatically update the incident commander's PC which, will give each company officer access to hazardous materials guidebooks, real-time weather, technical information on various types of hazardous materials, and much more. The software automatically logs all communication between the sectors and the commander building a real-time activity log. Although this author has no financial interest in the company, mention is made here of the system because of its potential for assisting those managing emergency scenes. Interested parties can contact the civil defense group and ask about their Cobra software. Although they appear to be a front runner, other companies will certainly follow their lead.

chapter eight

Eliminating, minimizing, and shifting risks

> "All business proceeds on beliefs, on judgments of probabilities, and not on certainties."
>
> Charles William Eliot

> "Show me a person who isn't taking risks. I'll show you a person who isn't running a business."
>
> Allen E. Paulson

In preparing for an emergency or disaster situation, prudent safety professionals may want to assess the level of insurance protection of the physical assets of the facility or operations. Although the specific responsibilities of acquisition and maintenance of specific insurance policies usually falls within the domain of the risk manager or insurance administrator, it is important for the safety professional to possess an adequate knowledge base of the levels of protections, benefits to be provided, and contact names for each insurance carrier to be utilized as part of the overall assessment as well as use in an emergency situation.

In most facilities or operations, insurance protections (i.e., insurance policies) are acquired to provide recovery in the event of a disaster situation. Insurance policies are usually written to protect the facility structure and specific contents. Additionally, insurance is usually acquired to cover vehicles, such as trucks, as well as general liability insurance. Workers' compensation coverage usually provides the coverage for work-related injuries or illnesses. Other insurances, such as directors' and officers' insurance and sexual harassment insurance, are also utilized by companies to offer specific protections.

In today's society, insurance can be purchased to protect virtually anything from Joe Nameth's knees to the Exxon Valdez. The primary issue for most organizations is the cost of the insurance and the level of protection provided. For example, if a 20-year-old male college student with a DUI and one minor car accident wished to acquire full coverage on a new Corvette, the insurance company would assess their risk of loss of the 20-year-old demolishing the car and provide a premium that would be extremely high because the risk is extremely high. The same concept holds true with property, casualty, and other insurances. If the risk of loss is low, the premiums correlate to the risk. If the risk is high, the premiums are high.

Another area of consideration in the assessment is the amount of the deductible. For most facilities, the amount of the potential loss could be in the millions, thus many organizations opt for a larger deductible in order to reduce the amount of the periodic premium of the policy. As an example, the cost of full coverage on the 20-year-old's Corvette would be $1000 per month. The cost of full coverage with a $500 deductible may be $900 per month; the cost with a $1000 deductible may be $750 per month. If the individual selects the $1000 deductible and has an accident, the first $1000 of the damage would be paid by the individual. With a policy for a facility valued at 10 million dollars, the deductible on the policy may be $500,000 or even 1 million dollars.

Given the substantial amount of many deductibles, safety professionals may wish to include these potential losses within their overall assessment. Additionally, items such as lost business, lost products, efficacy losses, and other potential losses not covered under any insurance policy may also be included in the overall assessment.

The primary purpose of inclusion of insurance protections in the overall emergency and disaster preparedness assessment is two-fold: first, to ensure that all risks have been identified and appropriate risks are protected through the acquisition of appropriate insurance protection; second, to assess the level of protection provided by the insurance protection to ensure that the protection provided is adequate and appropriate to the level of risk present.

To assist the safety professional in assessing the primary risks usually present in most operations the assessment checklist follows:

1. Is the physical structure insured? Yes No
2. Is the level of insurance appropriate? Yes No
3. Is the equipment inside insured? Yes No
4. Is the level of insurance appropriate? Yes No
5. Is all new equipment protected? Yes No
6. Are all vehicles insured? Yes No
7. Do you have workers' compensation? Yes No
8. Do you have general liability insurance? Yes No
9. Do you have insurance for third-party actions? Yes No
10. Does your company indemnify your officers? Yes No
11. Are there any risks not provided protection? Yes No
12. Are there any risks that should be protected? Yes No
13. Do you possess explosive hazards? Yes No

14. Does your facility possess natural hazards?   Yes No
15. Is specialized equipment/individuals protected? Yes No
16. Are business losses protected?   Yes No
17. Are areas of litigation protected?   Yes No
18. Is there potential of malpractice?   Yes No
19. Are areas of major expenditures protected?   Yes No
20. Are company trade secrets protected?   Yes No

Safety professionals work daily in the development of programs and training of personnel to minimize the potential risk in the workplace. As with the overall efforts of safety professionals, other risks in the workplace, such as fire, explosive, or water damages, can be minimized through proactive efforts and programs to minimize the risk of harm. Such programs are usually specialized to the specific risks in the workplace, but the basic principles utilized by safety professionals in their daily efforts usually provide a solid foundation from which to initiate these efforts.

Wherever possible, the potential risk of harm should be removed from the facility or operation. In conducting the overall assessment, safety professionals should look for areas of risk that can be eliminated or substituted from the operation. For example, a specific chemical which is carcinogenic is being utilized in the operations. The safety professional should search for a substitute chemical that can perform the function, but does not possess potentially detrimental long-term health effects. Additionally, as often found in maintenance areas, certain chemicals are used on a very infrequent basis, but the chemicals are very dangerous. If a dangerous chemical is not being used, eliminate it from the operations.

And lastly, if the potential risk cannot be eliminated or minimized, prudent safety professionals may want to consider the acquisition of some level of insurance protection to reduce the detrimental impact on the overall operation in the event of a disaster situation. As always, an appropriate risk v. benefit assessment should be included in the overall assessment to ensure the level of risk is appropriate and the level of protection provided meets the specific needs of the situation. If the risk is adequately high, a percentage of the risk may want to be shifted to an insurance carrier or other insurance entity.

In summation, insurance companies are in business to make money. If you pay your car insurance premium every month and do not have an accident, the insurance company makes money. If you have several accidents and a few DUIs, your risk of loss has increased considerably and either your premiums go up or you are dropped by your insurance carrier. In most organizations, substantial efforts are made to protect the physical and human assets of the company and reduce the potential risks. However, where the potential risk of loss is substantially great and the risk cannot be eliminated or minimized to an appropriate level, shifting the risk of loss, in whole or in part, may be a consideration for the organization.

chapter nine

Developing an action plan

> "By failing to prepare you are preparing to fail."
>
> Benjamin Franklin

> "The executive of the future will be rated by his ability to anticipate his problems rather than to meet them as they come."
>
> Howard Coonley

You have assessed the risks, probabilities, vulnerabilities, identified the resources, and you have started to pre-plan individual hazards and potential emergencies. If you have not already done so, it is time to develop an action plan for writing this emergency plan and implementing it. Without a doubt getting a strong commitment from upper management is essential to obtaining the cooperation and support of the employees. Communication about the formation of the emergency planning should begin to appear in company newsletters, bulletins, etc. Once the emergency planning team has been selected (more about member selection in Chapter 12), the highest ranking official in the organization should introduce the team and discuss his support of their work and its importance to the future of the company. This should be accomplished in a formal meeting that demonstrates the management commitment.

Based upon the capabilities of surrounding agencies, vulnerability to disasters, willingness of your own personnel to engage in fire, EMS and Haz-Mat responses, and management commitment to training and equipping a brigade, a mission statement should be developed that serves as the basis for the entire emergency plan. It should detail the purpose of the plan and the total commitment of the company to implementing it. If your employees will do more than evacuate the premises, the response brigade you are creating should also be included in the mission statement. The mission statement should:

- Demonstrate management support and commitment to the planning process
- Give authority for the planning group
- Define the structure of the planning group
- Define the authority and structure of the emergency response group or brigade
- Detail those activities which the response group will be trained and equipped to handle
- Provide a realistic timeline for implementation
- Be signed by the Chief Executive Officer

The plant manager can lead the initial planning but at some point an emergency planning director or manager should be appointed by management. This individual should manage the daily activities of developing and implementing the plan. He or she would also be the company point of contact or liaison with outside agencies for emergency planning. This individual would also be able to answer any questions regulatory or response agencies might have about the plan. In the event of an emergency, this person could command the in-plant response and coordinate support for outside agencies as they arrive.

The team should begin developing a draft document describing tactical expectations during various types of emergencies. This document details exactly what brigade members will be authorized to do, the conditions, and any limitations. Initially, tasks may be limited to duties such as: emergency notification, employee evacuation, internal communication, and external communication with emergency response agencies. As the brigade receives training, equipment, and confidence, duties such as manual fire protection, hazardous materials response, and confined space rescue may be added if the mission statement authorizes them. The outside agencies must be informed of any increasing roles of the in-house brigade, and joint training should be arranged to facilitate working together in emergency situations.

Next, meetings with outside agency personnel should begin to acquaint their personnel with your hazards and begin to develop strategies for dealing with those hazards. Any special equipment that is essential to the success of the plan should be budgeted and purchased as quickly as possible to permit training with the equipment. Hazards should be prioritized based upon their potential negative impact on life, property, continuity of operations, and community support. If a particular hazard has the potential to kill, injure, or destroy the company, it must be dealt with first. All hazards that can impact critical systems must also be addressed quickly.

Although good emergency planning seeks more than just regulatory compliance, the planning team must study OSHA, EPA, SARA Title III, and other local, state, and federal regulations to ensure compliance with their emergency planning and reporting requirements. As draft copies of the written plan emerge, meetings should be held with all responding agencies to obtain their buy in and ownership of the plan. Any agencies responding outside their

Chapter nine: Developing an action plan

jurisdictional boundaries should notify their insurance carriers of these agreements to ensure coverage for their vehicles and personnel. Critical elements of the plan should become the first training and testing exercises. It is much better to find a problem with a critical aspect of the plan before a real emergency happens. After the draft documents have been reviewed by all parties and approved, the documents should be formalized by signing agreements committing to the plan. This will also provide authority and protection for agencies responding across traditional boundaries.

The emergency plan should never be finished completely. As the response to the worst hazards are polished to perfection, lesser hazards can be refined. Changes in construction, production, personnel, and process will demand constant updates to the plan. It is imperative that each page of the plan be numbered and each section be dated to enable the emergency plan manager to furnish everyone with the latest copy and dispose of outdated plans. Again, the emphasis on position titles in the document rather than personnel names will allow replacement of an outdated phone list much more easily than constantly reprinting pages of the plan.

·itten plan

"will not make his boat a

Amenemhet I

plan!"

Anonymous

you to develop emergency response
y can only use one plan to prepare,
that occur. Your challenge is to study
al regulations and produce a single
Fortunately there are some elements
ould include the following:

- An introductory section with:
  - A statement describing the purpose of the plan and management's commitment to emergency preparedness
  - A statement as to whether the facility will evacuate only
  - A statement as to whether the facility will have some form of active emergency response, the mission, and scope of that brigade
  - The name and contact information of the individual to be contacted concerning questions about the plan
- The body of the document that includes at least:
  - A flow chart detailing the chain of command by position titles for each type of emergency (violence, bomb threat, medical emergency, fire, Haz-Mat release)
  - A site map or series of maps detailing important emergency information
  - An emergency contact list of all involved employees
  - An emergency contact list of all agencies that participated in the plan and have agreed to respond in the event of an emergency

- A list of communications frequencies that various responding agencies have available (ideally all agencies would have a common set of frequencies available)
- A resource contact list of agencies and companies which may be needed in the event of an emergency
- A description of each of the different types of hazards and emergency responses that were identified in the risk and vulnerability analysis
- The proper action an employee should take when confronted with each different type of emergency
- A statement describing the authority and responsibility of various agencies at each type of response identified

It is at this point that emergency response plans begin to differ from one company to another. Some companies prefer to provide detailed response guides to each specific hazard in the main emergency response plan. Other companies prefer to write very broad response guidelines to the various types of emergencies (fire, medical, hazardous materials, etc.) and refer the reader and response agencies to standard operating procedures (SOPs) developed for each target hazard identified. These tactical plans dictate duties and responsibilities for responding agencies either by specific company or in order of arrival on the scene. For example, a report of fire alarm sounding in Building C might have a line on the SOP that reads as follows:

> The first arriving engine company will set up command on the southwest corner of Building C, stretch a hose line from hydrant C12 to the fire department connection, and await orders from command to charge the connection.

One advantage of these SOPs is their ability to be used for training. These SOPs should be designed around the priorities of life safety, property protection, continuity of operations, and protection of the environment.

Other companies have utilized a hot book–cold book combination. The cold book printed on blue paper at the back of the emergency plan contains all sorts of regulatory compliance information and introductory material and possibly specific guidelines for identified hazards. A much smaller book attached to the front of the plan is printed in larger text with a flow chart on red or pink paper. The flow chart asks the employee to answer very simple questions about the nature of the potential emergency. It then either gives the employee specific response instructions, such as activate the evacuation alarm and exit, or refers the employee to the appropriate section of the blue or cold book for additional information, such as the location of the nearest alarm station or details about the evacuation plan. The emergency plan must be easy to use under emergency situations. An easy to read table of contents and/or index is another way for employees to find the information quickly. Another idea that has enjoyed success is color coding various sections of the emergency plan to correspond to different types of emergencies. For example, red paper might be used for fire response information, blue paper might be used for medical emergencies, and yellow paper could contain information about hazardous material responses.

The goal of the plan is for every employee to be able to determine what to do and where to go in an emergency. It also provides additional information for response agencies. Once the plan is written in draft form, it must be submitted to the entire team for editing and corrections. It is common for the response planning team to assign various sections and hazards to sub groups of the team. This team review and critique will yield valuable information. After these corrections and changes are incorporated, the document should be tested in a drill with a simulated emergency. Any problems discovered using the document or the related SOPs should be corrected. Remember to test all types of emergencies. This will test the unified command structure, which places the agency with the most expertise in charge and allows the other agencies to serve in support roles. For example, a crime scene or violent incident involving guns should be directed by law enforcement, and a mass casualty incident would be better directed by EMS or possibly fire commanders.

After the plan has been sufficiently tested and any modifications or corrections made, the plan should be printed and distributed to all response agencies and employees. All employees should be trained in the proper use of the manual and if an internal response brigade exists, they should begin to train with the appropriate sections. Creating awareness of and ability to use the emergency response manual should be incorporated into the initial training requirements of every employee hired.

chapter eleven

Effective communication

"Men are never so likely to settle a question rightly as when they discuss it freely."

Thomas B. Macaulay

"The most valuable of talents is that of never using two words when one will do."

Thomas Jefferson

A universally accepted system of management and command

Once you have a commitment from the initial response agencies, begin to assess their capability to work together, train together, and communicate over radios, telephone, etc. The ability to effectively command the incident using a unified command structure is a key to successfully handling any emergency. The incident management system (IMS) also known as the incident command system (ICS) or sometimes the fire ground command system (FGCS) is a modular system designed to grow as an incident escalates enabling those in charge to maintain a reasonable span of control. In an emergency situation, an officer can only effectively manage 3 to 7 people. As the first unit arrives, the company officer is in charge until a superior officer takes over. If the first arriving units are overwhelmed by life safety issues (rescues), the company officer may delay setting up a formal command post by passing command to the next arriving company officer. Because the first company on the scene is under his control, the officer still has an effective command structure in place even though a formal command center has not been established yet. As the emergency progresses, additional resources will be deployed and divisions, groups, or sectors will be established, each with its own officer. Each time another layer of command is added, the opportunity to transfer command to another higher ranking officer exists. The commanders are building a giant pyramid that enables each officer to only have to interact with 3 to 7 people. In very large scale incidents, five functional positions are allocated:

- Command
- Operations
- Planning
- Logistics
- Finance

Command is the function directing the overall incident through the incident commander (IC). This function is always staffed even if a single company is operating. If the incident progresses beyond a few companies, the IC frequently creates command staff officer positions for a safety officer (SO) and a liaison officer (LO). In large scale incidents it is also wise to quickly establish a public information officer (PIO) who constantly records events as they happen. A Littleton, CO, the fire officer described the time consuming task of listening to hours of radio transmission recordings to reconstruct the exact time and sequence of events that followed the April 1999 Columbine High School shootings. The public information officer also communicates with the media and is usually assigned to the command staff.

During the Los Angeles riots which followed the Rodney King police brutality trial verdict, the Los Angeles Fire Department PIO was able to explain fire department tactics to the television reporters and turn around negative reports of the firefighters' actions. The reporters will report what they see and perceive. It's up to you to see that they have accurate information. Since the PIO, SO, and LO are all resources and advisors for the IC, they are not considered and counted in the IC's area of control.

Operations is the area responsible for implementing the tactical objectives of the incident commander. The operations commander works with the group, division, and/or sector officers who are attempting to mitigate the emergency. They may be ventilating the roof, searching for victims, staffing hose lines, containing spills, or other tactical objectives.

Planning is the position that gathers information and analyzes it to forecast the impact of the current plan of action and make modifications as necessary for a successful operation.

Logistics is the function that ensures that resources continue to be available as needed. Items such as fuel, food, medical services, specialty equipment, additional vehicles, and personnel are examples of support that must be provided if the tactical operations are to continue. You can imagine the size and scope of the logistics operation in the Oklahoma City bombing.

Finance is the function that is usually only staffed at major incidents. Large-scale operations require fiscal documentation of expenditures, and the finance officer may also assist the IC with financial planning and regulatory issues. If negligence is later determined to have caused the emergency, the finance officer's documentation of expenditures may help the department recover some of the cost of operations. This cost recovery is becoming increasingly popular in hazardous material releases.

As long as everyone on the scene practices and operates within the ICS system, the operation progresses smoothly. Occasionally, some responding mutual aid companies may slip through the staging process and arrive on the scene in the *freelance mode*. This type of activity is dangerous and may adversely effect an otherwise safe and sound response plan. A classic example of this involves freelancers directing hose streams into openings cut to remove smoke and heat. The crews on the inside of the structure advancing hose lines to the seat of the fire will have smoke, heat, and steam forced back on them resulting in their retreat and/or injury.

Radios, telephones, and emergency operations centers

The task of managing communication at a large-scale emergency scene is not always an easy one. In an ideal world, the incident takes place in an area where all responding agencies are capable of sharing common radio frequencies. There would also be an endless supply of fully charged batteries for all the portable radios. Police, fire, EMS, and public works agencies do not routinely have to talk to each other, but at certain incidents the ability to determine if the person is supposed to be in that location can mean the difference between life and death. The actual emergency operations center (EOC) rarely has to be located on the property of the incident. Information can be relayed via radio, cellular telephone, fax, and digital imaging to remote sites. Communication vehicle personnel can set up mobile communications command and assist the operations commander by assigning talk groups to various divisions or groups. This can minimize radio chatter. Communications command can also facilitate frequency switching for commanders who need to talk to people on other channels. The EOC can be established in a communications vehicle or building near the emergency site, but frequently the mission of information processing (receiving, relaying, planning, logistics, finance, and other duties) can be accomplished in a location remote to the incident. Incoming telephone calls can be screened and directed to the proper individuals or, if not of an emergency related nature, politely terminated. This frees up people and telephone resources for essential communication.

As is often the case, responding agencies may operate on different radio frequencies. This must be determined early in the planning phase and a sufficient number of mobile and portable radios must be purchased to enable all units to communicate directly or through relays of transmissions to other units. This is especially important at scenes involving violent criminal acts. Police officers must know if the individuals in their sights are the good guys or the bad guys. Inability to communicate may allow dangerous individuals to escape, take additional hostages, or kill and injure more people. Fire and EMS units must be able to summon help and search for victims without fear of being shot by their fellow public servants. Some bombings and shooting incidents have reported over 500 law enforcement officers on the scene in hours. Command and control of that many people must be managed under the ICS and effective communication capability is imperative to prevent misunderstandings, tactical blunders and accidental injuries and deaths.

In the event the situation takes days or weeks to resolve, portable radios, cell phones, and many more items will have to be used around the clock with sufficient batteries, chargers, and spare radios and telephones available as problems arise with existing equipment.

Internal communications

Fire detection and alarm systems must be inspected, tested, and thoroughly understood by both maintenance and emergency team members. False alarms tax the resources and patience of in-plant personnel and outside response agencies. The proper selection, installation, maintenance, and testing of fire alarms is the first step in minimizing accidental activation. The NFPA 72 series standards provide information concerning fire alarms. The ability to interpret coded fire alarm signals and respond to the scene quickly can allow team members to handle a fire in its incipient or beginning stages with a portable fire extinguisher before the fire has grown to the size requiring a sprinkler head to activate. Immediate response and investigation of alarms can often turn back responding units minimizing their time out of service and, more importantly, minimizing their risk of being involved in a vehicle accident. Cleaning of fire detectors is a commonly overlooked activity that leads to false alarms. Maintenance personnel with portable radios, cellular telephones, or pagers can be dispatched quickly to size-up the situation and decide what resources, if any, will be needed. It is worth repeating that when an alarm panel indicates a valve is open or a fire pump is running, do not believe the indicator. If a water flow alarm is sounding or a fire pump is running, you will need assistance to clean up water at the very least. Activate the emergency response system and get help on the way. In addition to sending someone to the incident scene, make sure that you send someone to the pump room, the sprinkler riser that is indicated, and to the fire control room or command station. As soon as the runners reach their locations, have them report the conditions to the incident commander and await further instructions. **Do not allow anyone other than your emergency team members to turn off or close any fire protection devices or valves without permission from the incident commander.**

Every employee should be instructed on the location and operation of the manual fire pull stations and/or other means of reporting alarms and other emergencies. If a single system is used for multiple notification purposes (i.e., paging alerts, end of shift signal, tornado, fire alarm, etc.) the emergency signaling tones must be unique and different for each type of evacuation. If hearing, vision, or other physically challenged individuals are in the facility, appropriate means for notifying them must be present. High intensity emergency strobes and personal vibrating pagers are among the means currently being used. These methods can also be effective in alerting employees in high noise areas.

Suitable building site maps with emergency routes must be available. They should be prominently displayed throughout the facility directing employees to the nearest facility appropriate for the type of emergency signal sounding. Separate maps can be used for fire, tornado, and whatever emergencies are anticipated, or the maps can be color coded for the various evacuation routes. Full scale evacuations take time, which can cost production dollars, but they are the only way of determining flaws in the egress signage, shelter capacity, employee accountability systems, and training of the employees.

Policies and procedures must also be developed for dealing with the possibility of workplace violence incidents. Some facilities choose to include the workplace violence policy as a part of the emergency plan and other facilities choose to treat it as a separate policy. Either way the emphasis must be on pro-active approaches to conflict resolution which prevent violent acts. The employees must be trained in the correct response to potentially violent encounters. Although the frequency of these random violent acts is low, your employees must be prepared properly to protect themselves and others. Access control systems are increasingly seen as a tool in discouraging angry spouses and disgruntled former employees from gaining access to the plant or areas where they can cause violent acts. Coded information on each employee's photo identification card will only allow the employee into those areas he or she must access to accomplish his or her job. Controlled access doors can be attached to logging devices to record, date, and time stamp everyone who enters and leaves that particular room or gate. These access control card systems can also be used to quickly assess employee evacuation with portable card readers at the accountability sites either in-plant shelters or at outside check-in stations. These same photo identification cards can also be used by the IC to maintain accountability of personnel working the emergency.

chapter twelve

Selecting the right people

> "Credit to the fullest the good qualities to be found in others, even though they may far outshine yours."
>
> James Russell Lowell

> "Some men see things as they are and ask, 'Why?' I dream things that never were and ask 'Why not?'"
>
> Robert F. Kennedy

The job of selecting individuals to assist you in developing the emergency response plan and of staffing the response team is an important one. The exact number of people who comprise the team will depend on the size, scope, and complexity of the facility and its operations.

Assembling an emergency response planning team

A diverse team with representation from many different groups and areas within the plant is advantageous because it:

- Provides more assistance to accomplish the tasks required
- Involves more people who then have a vested interest in the plan's success
- Enhances problem solving by allowing people with different backgrounds, experience, and expertise to the view the problems from different viewpoints and paradigms
- Draws upon the expertise of those individuals who know the operations and processes best

Having a representative from upper management with overall advisory responsibility for developing and implementing the plan will show management commitment and provide the budgetary resources required. The chief executive officer should address the group and describe the importance of the plan and the management executive who will be responsible for guiding and implementing the project. Unless that same individual will be the emergency team leader and/or emergency plan coordinator, whoever fills that position and ultimately manages the plan should be in charge of developing it. This individual should lead the planning team. Some obvious choices for this position are the plant safety manager or, if the plant has an organized response brigade, the leader of that brigade. A plan may already exist and your task may be to bring it up to date.

All the different work units in the facility should be represented since each area will bring unique challenges to dealing with the emergency that may arise in that work unit. The line supervisors who manage the production workers should be represented. If the labor force is organized, they should be represented by their leadership, or representatives from the labor force can be solicited from the pool if the force is not represented by a union. Representatives from safety and environmental committees should serve. Marketing, public relations, engineering, maintenance, human relations, and legal counsel should also be represented. Purchasing, finance, security, and telecommunications areas should also have members on the planning committee. Adequate secretarial resources should be provided to support the planning team. These individuals will be needed to schedule meetings, and record and transcribe the minutes. These minutes will reflect action items identified, the names of the staff responsible for accomplishing them, and due dates. This will keep the team on track and discourage team members from neglecting their duties or tasks. Participation on the planning team should be by those individuals who want to serve. If a particular person representing a work unit has been delegated to serve and does not want to participate, the team leader should find a way for that person to exit gracefully and be replaced by a person who is willing to serve. When the team begins its work, it will be necessary for most of the team to attend meetings because input will be sought from the entire group to assist in determining the hazards present, their severity, the probability each may occur, and finally the company's vulnerability if they do occur.

Based upon their individual expertise, team members will be asked to contribute to various sections of the plan. As the plan begins to take shape and individual hazards are being pre-planned and SOPs developed, smaller diverse groups with specific expertise about the equipment, process, and control will work with response agencies to develop appropriate prevention and intervention techniques. To maintain a standard format and appearance, a small group will develop the individual hazard response guides or SOPs using a style sheet. Standardizing the appearance of the SOPs will enable emergency response personnel to learn where to look for specific pieces of information. As each SOP is developed it should be considered in draft form until it has been reviewed by engineering, maintenance, and safety. Emergency response personnel should be encouraged to practice or drill with the

Chapter twelve: Selecting the right people

SOP until they are confident they have identified and corrected any potential errors or problems with that SOP.

The team working on emergency contact lists and emergency resource lists should use the draft lists to contact everyone on the list validating the information. An emergency is not the time to find a bad telephone number or business address.

Other team members can compare the documents being developed to the regulatory requirements of local, state, and federal government. Local Emergency Planning Committee (LEPC) team members should be able to provide assistance in this area. Maintenance and engineering should verify the accuracy of any maps included in the plan. The company's insurance carrier employs loss control representatives who will provide expert technical assistance in developing your plan. They can also review the plan and make suggestions for improving it.

Selecting employees to serve on the emergency response team or brigade

As you begin to search for emergency response team members, keep in mind the activities your employees participate in when they are not working. Many of these people have unique skills which can be valuable to the team. Any employee who is a member of a fire department, rescue squad, or ambulance squad already has special knowledge, training, and experience that will translate to the emergency response team. Some may even have command experience or be certified to function at hazardous materials incidents. These people should already be medically qualified as being physically fit and capable of wearing respirators and protective clothing. They are familiar with emergency scenes and have already demonstrated their commitment to the community. Failure to at least invite these people would be a mistake. Even retired emergency service personnel can function as trainers and could possibly serve in advisory and staff functions with your response team. People with previous military experience could also transfer certain skills from those jobs into useful functions within your brigade. You should invite and solicit volunteers even if you intend to provide some kind of supplemental pay. In this manner you will be certain you have people who want to be on the team and not just increase their pay. You will be responsible for medical screening and surveillance to ensure that all active team members are physically capable of wearing respirators, protective clothing, and doing strenuous work. If you intend to use trainers and advisory or support staff who are not medically certified, you will have to take steps to prevent those individuals from participating in strenuous activities and real emergency scenes. These individuals can assist with scheduling training sessions and drills, conducting classroom training, tabletop exercises, and serve as evaluators or patients during full-scale disaster drills. Previous experience in emergency organizations should not be viewed as a prerequisite for being on the response brigade. Anyone who is physically fit, willing to learn, and interested in protecting the plant and their job is a potential member. Even if the person is uncomfortable wearing self-contained breathing apparatus, they can overcome their fears and be excellent assets to the organization. Although emergency response roles were traditionally viewed as suitable only for men, they are increasingly being filled by women. This author has trained many female industrial and municipal firefighters. Do not view these female team members as being any less capable than their male counterparts.

Once the team has been selected, evaluate the members' individual capabilities and develop a training plan that utilizes the skills of the more experienced and capable people to train and work with the newer people. Enlist the services of local firefighter training agencies, colleges, and universities. Develop enough training activities to keep the members interested without burdening them with the feeling they now have two jobs. Develop recognition and incentive programs to show these people they are appreciated. Annual recognition dinners, brigade member logo clothing, and letters of appreciation from upper management are among the incentives that other organizations have used. As the team's knowledge, skill, confidence, and equipment levels rise to meet and exceed OSHA and NFPA mandated training levels, revise the role your brigade plays in the overall response plan. The outside agencies will welcome the additional help and the reduced response time of your on-site team will catch many emergencies while they are small enough to minimize the threat to everyone.

chapter thirteen

Training for success

"They know enough who know how to learn."

Henry Adams

"The roots of education are bitter, but the fruit is sweet."

Aristotle

Safety professionals are usually responsible for conducting and/or managing the emergency and disaster preparedness training function in their organizations and companies. This important function is often *back burnered* by many safety professionals in anticipation that they can rely on videotape or other audiovisual aids. This tact is absolutely wrong and often results in a wasted opportunity, and can send the wrong unspoken message to your employees and management team members regarding emergency and disaster preparedness.

Training and education programs provide the safety professional a unique opportunity to transfer important information to employees and management team members while also providing an opportunity to interact and familiarize himself/herself with the employees and management team members in a more relaxed educational environment. Preparation and presentation are the key elements in providing an effective and motivating training and education session. This interaction also provides the safety professional with an opportunity to identify individuals who may possess skills and abilities that would benefit your overall program while also identifying potential *weak links* in your system.

Safety professionals are often confronted with a wide variety of different types of training and education programs for their management team and employees in the overall emergency and disaster preparedness program. In most organizations or companies, the range of training can vary from new employee orientation training through education of the board of directors in the roles, responsibilities, risks, and liabilities in the area of emergency and disaster preparedness. Safety professionals should always properly prepare prior to initiating any training and ensure that all materials, curriculum, audiovisual (AV) equipment, and every other aspect of the training is choreographed to ensure maximum efficiency and complete understanding by the receiving parties.

One of most important types of training conducted in most organizations and companies is that of a required government agency compliance training. This type of training possesses very specific program elements which must be met as well as documentation requirements in order to prove this training has been conducted in accordance with requirements. Compliance training can include, but not be limited to, training for specific occupational safety and health standards (i.e., emergency and disaster preparedness).

In many organizations, training and education programs to inform as well as motivate employees and management team members are offered. In many cases, this training may provide specific methodologies as to improvement of management skills and other basic skills in the various areas of emergency and disaster preparedness. Motivational training is often incorporated into a series of training programs at various levels for the improvement of performance, communications skills, and other areas.

In many organizations, new employees and/or employees changing departments or areas are required to participate in a new employee orientation and training programs, which encompass a large variety of topic areas ranging from where to acquire your paycheck to the safety and health rules. This is very important training given that it is the first chance the safety professional has to interact with the new employees and set the tone for the employees' future with your company. Additionally, it is important from **day one** that all employees know their emergency and disaster program responsibilities, evacuation routes, and other vital information.

With virtually every type of training and education program, some type of follow-up and monitoring is required in order to ensure that the information transmitted during the training program is being utilized by the employees or management team member. This training can also include compliance refresher training as required by some requirements as well as specific follow-up reinforcement training for management skills. Emergency and disaster preparedness training should be an annual event for all employees in your organization.

In the safety professional's training activities, he or she may be faced with a wide variety of educational levels through which to transfer important information. It is essential that the safety professional identify the educational level of the participants in the specific training program so that the materials and program can be adjusted to transmit the information to the participants at the most appropriate of level possible.

In many organizations, the safety professional is responsible for training and educating the upper management group, including but not limited to board of directors members, presidents, CEOs, plant managers, and others in the various aspects of emergency and disaster preparedness. These topic areas may include cost issues, evacuation routes, recent court decisions, OSHA standard changes, and other specific topic areas. Most upper management education and training programs are offered at an advanced learner or college level.

At the advanced level, training and education is normally focused on a specific topic and the information provided is very focused. Supplemental materials in this type of training can normally include full case studies, advance research information, full text of the law, and other unedited documents. Often with this type of training, a summary of the information should be provided to these busy participants as well as full text materials. Use of audiovisual materials are normally limited to very focused items and the training time provided is normally a shortened duration.

Safety professionals are often responsible for the development of education and training programs for their supervisors, team leaders, and other members of mid-level management for the emergency and disaster program. This training is often focused at the high school level; however, some organizations increase the educational level through college level depending upon the topic area and background of these supervisors or team leaders. This training can include a variety of areas from evacuation routes to specific emergency responsibilities.

In many organizations, the safety professional is responsible for the development of new employee orientation programs as well as ongoing employee training that normally is focused in the area of compliance. Employee training is normally developed at approximately an eighth to tenth grade level dependent upon the background of the work force.

The training cannot only be designed in accordance with the educational level of the participants, but also in different formats to maximize the level of understanding. Additionally, with the new technologies, training can now be provided via interactive computer programs or can be transmitted to the location via distance learning technologies. Several formats for consideration include the following:

- **Classroom (Formal) Training** — the traditional method of learning with an instructor providing information to the participants. This format can be provided in lecture and/or facilitating manner.
- **Hands-On Training** — especially in the area of compliance or job specific training, the training program can be focused around the employee actually performing the required functions under the supervision of a trained instructor or supervisor.
- **Computer Assisted Training** — depending on the level of technology available in the facility, computer assisted training utilizing compressed videotape and/or designed programs can be provided to employees.
- **Interactive Computer Training** — the newest technology permits the participant to interact with the computer program to acquire the necessary education level in testing within the software program.

The goal of virtually every level of training is to properly communicate the required information to the participant and ensure complete understanding of the information. Depending on the organization in the type of resources available, the safety professional can adapt the information and develop a training program to meet the needs of the specific participants for the various important emergency and disaster preparedness elements.

From the moment we are born, human beings have acquired information using our senses of sight, speech, smell, hearing, and touch. In the industrial world, the primary senses that we utilize to learn are sight, speech, hearing, and touch with sight and hearing being used most frequently.

Traditional learning was focused primarily on the senses of sight and hearing. However, safety professionals should explore additional avenues to increase the level of understanding and focus the training around specific subject matter. In today's society, we have become accustomed to watching a television and/or utilizing a computer screen to acquire information. In most adult learners, information is acquired primarily through sight and verbal explanation; although the attention span is greater, most adults require an entertainment factor to maintain a high level of concentration.

With specific technical skills, incorporating the use of touch can increase the level of understanding. As with riding a bicycle, we can easily tell and show an individual how to ride, but until he or she actually gets on the bicycle, the skill proficiency in riding a bicycle will not be maximized.

- **Visual Learning** — safety professionals can maximize the visual learning capabilities of adult learners through the use of videotapes, slides, overheads, and computer generated information.
- **Auditory Learning** — verbal information must be presented in a decibel level range (not to high or not to low) and in an entertaining manner to maintain the interest of an adult learner. Monotone voices, screechy voices, noise interference, and other distractions can affect auditory learning.
- **Learning Through Doing** — hands-on training has been found to be the most effective method, when combined with other types of training, to acquire proficient skills and employees. This additional use of the employees' sense of touch increases the learning curve. For example, walk the employees through their evacuation route.

The time provided by most organizations for education and training programs is normally limited especially in the area of emergency and disaster preparedness. Safety professionals should maximize this limited time in order to achieve the most effective results. Appropriately schedule training and education activities to maximize the learning of the employees or management team members. Safety professionals should effectively plan every aspect of the activity from the setting arrangement to the final testing. All curriculum, audiovisual aids, documents, supplemental material, and other required items must be prepared in advance to maximize efficiency.

In most organizations, the time provided for education and training programs is normally before the shift starts, immediately after the shift, at lunchtime, or other non-productive work hours. These are normally not the best time frames for learning for most individuals. Prior to work, the individuals are in a routine and are not normally fully prepared to learn. Following a work shift, the employees may be tired after a full day's work and may not be able to properly focus on the information being provided. During lunch time and breaks, employees are often busy eating lunch or taking care of other normal activities while the information is being provided. To maximize the participants' learning, the safety professional should attempt to schedule training and education programs during the maximum peak period for the individual. This will vary depending on the number of shifts (i.e., 3–11 p.m.; 11–7 a.m.), and the personnel and human resource manager should be flexible in scheduling training to achieve maximum results.

Every minute of an emergency and disaster preparedness training program is crucial and thus the maximum amount of information that the individual can absorb and retain should be utilized. To do this, the safety professional should properly schedule and design an agenda to be closely followed during the training program. Time should also be allotted for employees to ask questions or acquire clarification of information; however, training programs are not the time to voice complaints or discuss unrelated items.

Preparation of training materials can include audiovisual materials, written handout materials, training manuals, etc. In some circumstances, the safety professional may be able to acquire a *canned* program which is readily available on the market to supplement their presentation (such as a videotape of a disaster analysis). In most circumstances, the safety professional should create and prepare this information for presentation to the participants, to the employees, or to management team members. To do this, the safety professional should have access to the appropriate research materials which often include OSHA regulations, EPA regulations, etc. Sources that may be utilized by the participant or by the safety professional can include in-plant sources, local libraries, universities, or most recently the internet. Especially in the area of emergency and disaster preparedness, a substantial amount of the government regulations and supplemental information can be acquired through an internet source at your facility.

As noted above, the level in which the educational materials on emergency and disaster preparedness are presented to the employees or management team members is essential. As a general rule, educational materials should be designed for the lowest level of education within the specific group of employees or management team members. The employees and management team members at a higher educational level can understand the information at a lower level; however, the employees at a lower may not be able to understand the higher level of information.

Is it better to give out your materials before the training session? Is it better to hand out your materials during the training session? Is it better to give the materials following the training session? This would depend on the organization. Giving materials out prior to the training session often provides the employee or management team member a chance to review the materials prior to training; however, it has been found that the training material is often forgotten and/or has not been reviewed. Handing out written information during a training session will direct the employee or management team member's attention to the written material rather than the spoken communications. If materials are being handed out during the training session, it is often better to provide the employee or management team member a period of time to read the information prior to providing additional oral communication. Providing written information following the training session provides the employee or management team member the opportunity to refresh his or her memory following completion of the course and for use of reference materials. However, in many organizations written material gets deposited in the garbage can immediately after the oral presentation.

Especially in the area of emergency and disaster preparedness, the acquisition of feedback and follow-up after the completion of the training session is essential. Employees should be given an opportunity to provide feedback during the training session; however, it is often important to acquire their thoughts and ideas at a later date. Follow-up and reinforcement of the ideas are essential to the learning process and should be included in any training program.

One method that safety professionals can use to enhance the over-all learning of the employees and management team members is through the use of audiovisual aids. Audiovisual aids should be utilized only to supplement the training program and cannot normally be utilized for training in and of itself. Audiovisual aids can include, but are not limited to, transparencies, slides, videotape, computer animation, and computer-generated programs.

Safety professionals will find that there is a wide variety of canned programs at varying cost levels available. These programs are normally generic in nature and often need to be modified for plant specific activities and operations. A good source to look for these types of programs is the Best Safety Directory, professional magazines, and library sources.

With today's technology, there are a number of disk and CD ROM programs readily available in a generic format for use on a wide variety of topics. Additionally, specific software programs can be developed for use by plant employees and management team members in a wide variety of formats. Computer-generated and CD ROM programs are normally listed in the professional magazines and computer publications.

Although the initial cost tends to be high, there are a number of interactive video and interactive training programs currently on the market. These programs tend to be generic in nature, but provide the opportunity for employees and management team members to work through the program at their own pace on a personal computer.

In virtually all compliance areas, management skills area, and a number of other areas, there are generic, canned written materials in a wide variety of educational levels ranging from the comic book version through full text.

This information is normally copyrighted and thus a site permit or per piece cost would be required. This information is normally generic in nature and is often difficult to modify for individual plant situations.

Depending on time constraints, budgetary constraints, and the training topic, safety professionals may want to consider the wide variety of audiovisual aids and supplemental materials that are available on the market. Conversely, where budgetary constraints may limit the available outside resources, safety professionals may want to consider development of materials in house and use of a VCR to develop in house videotapes. Above all, safety professionals should utilize the resources available in an efficient manner and maximize the learning capabilities of the employees or management team members. Remember that simply putting a videotape in a VCR is not training. Use every possible source to maximize the efficiency of your training.

Just as with your emergency and disaster preparedness program as a whole, preparation is the key to success in any type of training and education program for employees or team members. Safety professionals should properly prepare the training materials including training schedules, curriculum, supplemental materials, and preparation for any training session. The training materials must be focused specifically on the participants' educational level and the specific training topic. The materials must be well written and presented in a manner in which the employee or management team member can absorb the information in a reasonable period of time.

Most safety professionals are bombarded by marketing materials offering commercially available training products ranging from training curriculums to on-site training programs. Again, as is noted above, these products are normally generic, but some can provide site-specific information if by the organization or company. Depending upon time and monetary constraints, these commercially available training programs may assist the personnel and human resource manager accomplishing his or her training needs.

In many circumstances, the safety professional cannot go to the open market and acquire a training program that is focused specifically on the needs of the individual employee or management team member in the area of emergency and disaster preparedness. Thus, the safety professional should prepare the training materials, course curriculum, supplemental materials, and other information in preparation for presenting the training program. Utilizing the "no stone unturned" theory, the safety professional should prepare the training program to encompass every facet from the preparation through the follow-up training to ensure the most efficient method of transmitting this important information.

Safety professionals should be aware that there are numerous sources that provide training information. This can include, but is not limited to, existing training manuals, videotape training courses, government publications, and OSHA standards. Additionally, safety professionals should inquire as to the availability of education and training assistance through local universities, state planning programs, and federal and state agencies.

Safety professionals should also explore possible sources of outside funding to assist in the training and educational endeavors. Often, there are federal, state, and even local programs that provide monetary assistance to companies and organizations for specific training and educational purposes.

In this age of the computer, safety professionals should be aware that there are a number of internet and web sources through which to acquire training and educational materials. These internet sources often provide a quick link to major educational data banks and provide specific information in the area of emergency and disaster preparedness.

Safety professionals often focus a substantial portion of their attention on training to ensure compliance with specific governmental regulations. These required training programs are often prescribed in the specific governmental regulation. Each element is set forth in the regulation or standard, and must be encompassed within the training program. Thus, governmental compliance programs must follow a specifically prescribed set of elements in order to achieve compliance. If a specific training requirement or element is omitted in the training program, the training program is not in compliance and thus can be cited as being deficient.

Safety professionals must be familiar with the wide variety of government regulations that require a training component including emergency and disaster preparedness. To do this, the safety professional must continually evaluate the laws and regulations with regards to the programs that are currently in place and be able to identify new or modified regulations or standards which have been promulgated.

Safety professionals may be able to keep abreast with the current changes and requirements in the training area through review of professional magazines, internet sources, and specific publications focused on identifying these issues.

Under most governmental standards, specific training requirements are set forth in published form. Safety professionals must be aware that specific adherence to the guidelines and standards is essential to insuring compliance.

In all truthfulness, the subject matter and/or program requirements of emergency and disaster preparedness can often be regimented and dull. Safety professionals should search for methods through which to present this material in an uplifting and entertaining manner while also achieving the compliance requirements. This can often be accomplished through the use of audiovisual aids and interaction between the employees, management team members, and the instructor.

It is essential that the safety professional document not only an employee's participation but also his or her understanding of the information provided in required training programs. If inspected, the compliance officer normally evaluates not only the written training program but also the documentation to prove that employees actively participated in and understood the information provided in the training programs. Absent this documentation, training programs are often found to be deficient.

Chapter thirteen: Training for success

Remember, the key elements of training for emergency and disaster preparedness program purposes include:

- Identifying the training requirements
- Preparing materials to meet governmental requirements
- Presenting materials to achieve maximum efficiency
- Documentation of governmental required training

Documentation of emergency and disaster preparedness training efforts by the safety professional is essential. The burden of proving that an employee participated and understood the information being transmitted during a training session falls upon the company or organization and thus the safety professional. Especially with required compliance training programs, it is essential that the safety professional document the training and understanding by the employee or management team member to the point where the training documentation will sustain scrutiny.

Traditionally, the method for documenting training education programs was for an employee or management team member to affix his or her signature to a *sign in* document. This document was normally headed by the name of the training program, the date, the instructor's name, and possibly the course curriculum. This method continues to be effective except in unusual circumstances such as the following:

- The training session is presented in English and the employee cannot speak or read English.
- The employee denies his signature and attendance in the training program.
- The employee indicates that he or she attended but did not understand the training.

To ensure appropriate documentation of training programs, several companies have initiated videotaping of the training session as well as more extensive documentation including, but not limited to, fingerprinting. To ensure that the employee understands the information, most companies are using some type of testing procedures.

How can you prove that an employee or management team member who participated in a training session really understands the information? This is especially important in the area of emergency and disaster preparedness in which an employee or management team member may allege that he or she participated in the training but did not understand the information, which resulted in a work-related injury, illness, or even a fatality. To ensure an appropriate level of understanding, many organizations and companies have initiated written testing or oral examinations to verify the employee or management team member's level of understanding. In most circumstances, this is accomplished through a written multiple choice or essay examination.

However, with specialized situations where the employee cannot read or write English or may have other difficulties, oral examinations are often utilized to accomplish this purpose.

In many organizations, the work force consists of employees and management team members who may speak or read a wide variety of languages. In these situations, safety professionals often develop specific training programs in the native language of the employees. If this is not feasible, often an interpreter is utilized in the training program and the information is provided in a multiple language format.

Documentation to verify that an employee or management team member participated and fully understood the information provided in a training program is essential. Several companies and organizations have avoided the *liars contest* in a court of law or before government agencies through the presentation of the signed training documents. The liars contest is when an employee denies ever receiving the training or understanding the training that was provided, which ultimately resulted in his or her injury, illness, or other problem. The safety professional, as the agent of the employer, provides oral testimony that the employee or management team member was properly trained but provides no documentation. Absent documentation, this issue would go to the jury or the trier of fact to decide who is lying in the particular situation. To avoid this, it is essential that the safety professional acquire and maintain appropriate documentation to be able to prove that the training was appropriate and was provided to the employee or management team member.

In many circumstances, the safety professional will need to utilize the documentation after a disaster situation has happened. Safety professionals do not know exactly when or where a particular event will happen; however, they should maintain the training documentation for a substantial period of time. The general rule of thumb is to maintain this important documentation for the life of the employee plus 20 years.

Documentation is essential to be able to prove that an employee participated in the required training program; however, improper documentation or falsified documentation can be a *smoking gun* for the safety professional. To this end, there are several general rules that should be followed:

- *Never* falsify training documents.
- Always require the employee or management team member to sign for himself or herself.
- Always maintain the documentation in a secure manner.
- Never have employees or management team members sign a blank sheet.
- Always keep your training curriculum, agenda, audiovisual aids, and other documentation in a secured place.

When training in emergency and disaster preparedness is interesting and fun, employees and team members are more apt to learn. We have all sat through training sessions that were extremely boring or were presented in an uninteresting manner. As safety professionals, we must capture the employee or management team member's attention and hold that attention with regard to the very important issues being presented for a period of time. To do this, if we present the information in an informative, entertaining, or even a fun manner, the employees or management team members are apt to maintain their level of attention and understanding.

It's very simple: keep your training upbeat and moving. Present the information in a logical manner and at the pace in which the employees or management team members must strive to maintain but is not beyond their comprehension. Safety professionals do not want their program to be paced too slow nor do you want it to be too fast. Like the story of the three bears, you want to strive to make your pace *just right* for each individual group. Strive to achieve the optimum level of understanding of the individual group of employees or team members in which the training is being presented.

There is nothing wrong with having fun even with a serious issue such as emergency and disaster preparedness. Provide exercises, play games, and provide the experiences that have happened in your plant. In essence, keep it interesting and keep it moving.

Audiovisual aids must be a supplement to the training and are not the "main course." Audiovisual aids should be utilized as appetizers to pique employees interest and questions. Always stress employees' questions and never embarrass an employee during any training or educational program. Remember, the training and educational program is for them to understand the information.

Most training and educational programs in the industrial setting are usually in blocks of one hour or three hours. Most employees and management team members possess a concentration level that can be maintained during these periods of time. However, depending on the time of day, day of the week, and other influences, the concentration level of the employee or management team member may be less than expected. Safety professionals should attempt to schedule training and educational programs at the optimal time and provide every opportunity for the employee or management team member to concentrate appropriately.

Your emergency and disaster preparedness training program should not be offered in a vacuum. After the formalized education and training program, safety professionals should provide feedback to participants, to employees and management team members, as to the status of the particular program or new information that may be acquired. This can be provided through postings on the employee bulletin board, use of graphs and charts monitoring progress, and other feedback mechanisms. Often, the formalized education and training program in the classroom is followed with actual hands-on instruction on the job. Many safety professionals have found that this combination of classroom and hands-on training provides the most efficient method through which employees learn. However, this combination of training can be intensive and costly, and it is important that the employee or management team member be provided appropriate mechanisms through which to absorb the information in an efficient manner.

Safety professionals who wish to incorporate hands-on training into their overall training program should provide the appropriate time, personnel, and availability of the particular machinery or apparatus to ensure maximum efficiency and on-the-job training time while on the shop floor. There is nothing worse than scheduling training and having the employees stand and watch another employee perform the particular function. For maximum efficiency, the employee must be able to touch, feel, and actually perform the prescribed operations.

With many hands-on emergency and disaster preparedness training activities, it is essential that the employee, the instructor, and the other employees and management team members in the area are safeguarded while the specific employee or management team member performs the procedure or job function. All appropriate safeguards should be in place and the instructor should be on a one-on-one basis with the employee while the specific job function is performed. All personal protective equipment, machine guards, and other safeguards must be in place prior to initiating this training.

Safety professionals should take advantage of every opportunity to expand the knowledge base and expertise of your employees and management team members in the area of emergency and disaster preparedness. Effective training and educational programs are a tried and true way to achieve the expected growth and competency in your work force while providing a method to monitor and evaluate performance.

Preparation and performance are the keys to a successful emergency and disaster preparedness training endeavor. Safety professionals should allot appropriate time, effort, and resources in preparing each and every training and educational session. Appropriate preparation will increase the confidence level of the safety professional and permit the presentation to be successful as well as fun.

chapter fourteen

Media control

> "If television encouraged us to work as much as it encourages us to do everything else, we could better afford to buy more of everything it advertises."
>
> Cullen Hightower

> "What the mass media offers is not popular art, but entertainment which is intended to be consumed like food, forgotten, and replaced by a new dish."
>
> W.H. Auden

> "People who complain our press is biased should note that during World War II the press was on our side — and we won!"
>
> Cullen Hightower

One essential area often overlooked in the preparation of an emergency and disaster preparedness plan is the control of the information and image being transferred to the world through the media about your organization and company. Pre-planning with regards to the who, what, when, where, and how of the information flow is essential to ensure the accuracy of the information being disseminated regarding your company and the emergency situation as well as the image that the public is making about your company through a 30 second sound bite.

Consider the following example: a publicly held company incurs an explosion that results in extensive damage to the facility, ten fatalities, and a large number of injured workers. Upon notification to the fire department, EMS, and local law enforcement, the local media, who is usually monitoring radio transmissions, will dispatch a reporter or television crew to the scene. The television crew will be working under a deadline to provide videotape and information regarding the incident as quickly as possible and within the specific time allotment available in the newscast. The videotape should be as graphic as possible to interest the viewers, and the information is acquired from by-standers, employees, firefighters, or whomever is available. The information may or may not be correct; however, the television station, for legal reasons, will add a disclaimer at the end of the information.

The information gathered at the incident scene will very quickly be acquired by CNN and other global television stations, be placed on the internet, and be on the shelves in newspapers. The information acquired may be slanted or otherwise *editorialized* and often, like telling a secret around a circle of children, begins to be modified, expanded, and otherwise changed to enhance the story. The facts and truth of the situation are often lost in the shuffle.

The viewer sitting at home watching the news or reading the newspaper is making a value judgment regarding the company or organization that incurred the disaster situation. This viewer may be a stockholder, a potential employee, a customer, or virtually anyone. This viewer is making a value judgment about your company or organization from a 30 second sound bite of videotape and commentary, which may have a bearing on his or her future employment with your company, purchase of your products, or purchase/sale of your stock. In essence, the information provided to the general public through television, internet, and other media is assisting in forming opinions in the minds of the general public about your company or organization — whether good or bad — which will affect their interaction with your company or organization in the future.

Control of the flow of information after a disaster situation is essential and should be part of your overall comprehensive emergency and disaster preparedness plan. In essence, it is as important today to control the information regarding the disaster situation as every other phase of the plan due to the long-term, down stream detrimental effects. As with all other components of the emergency and disaster preparedness plan, appropriate attention should be provided to all areas of the information flow to ensure that your company is putting the best "spin" on an already bad situation. Remember, a bell once sounded cannot be un-rung.

As part of your overall emergency and disaster preparedness efforts, please consider the following:

- Where will the media acquire their information?
- Who will be providing the media with the information?
- What image is being portrayed by the person providing the information?
- Where will the media park their vehicles?
- What background is behind the individual providing the information?
- What videotape footage will the media acquire?
- What does the media know about your company or organization **besides** the disaster situation?

Chapter fourteen: Media control

- Will the media detrimentally affect the emergency efforts?
- What will the individual providing information look like? (i.e., dress, voice quality, etc.)
- Is the individual providing the information a good representative of your company?
- Will information provided to the media be scripted or *ad-lib*?
- Will information be screened by legal counsel before being provided to the media?
- What is the time table for the media at location (i.e., 4:45 p.m. for the 6:00 p.m. news)?

Control of information is essential in order to minimize efficacy damage after a disaster situation. Consider the following measures to address the various media-related issues posed above as part of the overall emergency and disaster preparedness plan:

- A designated area in the parking area or away from the flow of emergency traffic to which security or management directs all media vehicles is identified in the emergency and disaster preparedness plan.
- Security is maintained in the media area to prohibit media representatives from wandering into the emergency area.
- A selected member of management is identified as the spokesperson for the company and no other members of management talk with the media.
- The selected spokesperson is provided an appropriate platform, microphone, and background (e.g., company logo) from which to provide information to the media.
- The dress, voice tone, ability to remain calm, and other attributes are considered in the selection of a spokesperson.
- The media is guided to appropriate areas to acquire video footage.
- Informational packets with company information are provided to the media.
- All information is screened by legal counsel prior to presentation and questions from the media are kept to a minimum.
- Always provide truthful information or no information at all.
- Remember the families of the injured or killed employees. Do not release names prior to notification of next of kin.
- Keep in mind the media's deadlines. Provide, if possible, information before their deadlines. If information is not provided, hearsay and file footage will be used.
- Remember, you're only news for a short period of time. This time is very intense and there is a lot going on at your facility or operation. You may be front page today and last page tomorrow. It is critical to control the information flow while the situation is most intense.

In conclusion, the media is a fact of life today. Properly managed, the disaster situation becomes just another blip on the information screens of the viewers. Improperly managed and the 30 second sound bite about the disaster situation can have long-range detrimental effects on your company that far outlast the disaster situation itself. And remember, everything said or viewed by the media is now locked on tape and possesses a high probability that the tape will be used in future litigation. Every aspect of the media should be controlled in order to put the best "spin" on a bad situation. Remember, a disaster situation just happened, emotions are high, individuals have been injured. Preparedness to handle the media in a calm, cool, and collected manner takes forethought and planning in order to properly manage the situation.

chapter fifteen

Shareholder factor

> "The market will not go up unless it goes up, nor will it go down unless it goes down, and it will stay the same unless it does either."
>
> Adam Smith

> "You don't buy stock because it has real value. You buy it because you feel there is always a greater fool down the street ready to pay more than you paid."
>
> Donald J. Stocking

Safety professionals should also consider the efficacy damage that a disaster could have on shareholder perception of the operations and on stock value in a publicly held company. Companies have become very bottom line oriented and the influence of Wall Street has become prevalent in day-to-day operations. A disaster of any magnitude will be communicated immediately through television, the internet, and other communication methods to the world. With publicly held companies being *owned* by the individuals and entities holding their stock, the value of the company is intrinsically tied to the values of its shares. If a shareholder perceives that the disaster will have a negative impact on the value of his stock, the shareholder will sell the stock. If a sufficient number of shares are sold, the value of the shares diminishes thus leading to a lower value for the company. Lower values equate to tighter budgets, fewer personnel, and other downsizing activity.

Safety professionals should be aware of the influence of Wall Street, which can have a positive or negative impact on important aspects of their program including, but not limited to, budgets and staffing. Additionally, with employees participating in 401K plans, stock, stock option plans, and other retirement programs, individual employees have become substantially more interested and are following the value of the company stock.

Let us examine this closely: if your company is public, the actual owners of your company would be the shareholders. Your company's board of directors actually works for the shareholders and your CEO and other upper management officials report to the board. The value of your company is based upon, in whole or in part, the value of your stock, which is traded daily on the New York Stock Exchange, American Stock Exchange, NASDAQ, or other stock exchange.

In addition to the number of individuals inside and outside of your company owning stock, technology has permitted individuals to track individual stocks on a daily basis through television programs such as MSNBC and through internet services. More individual shareholders are managing their portfolios through online trading such as E-Trade, Ameritrade, and other services. Previously, individuals and employees may have placed their money in a mutual fund and checked the progress once or twice a year. They are now monitoring the performance of individual company stocks on a daily or even a minute-by-minute basis. Individual shareholders are now more actively involved and more willing to purchase or sell an individual stock on a moment's notice rather than weather the storm.

So how does this directly and indirectly affect the overall function? First, upper management is now more focused on a quarter-by-quarter basis and looking for the return on investment or ROI for their shareholders. With this emphasis, the safety function is now placed under a microscope in order to minimize losses that may affect the bottom line. Next, given the instantaneous communication available today, large accidents, chemical releases, and other incidents will make the news immediately, which may have a detrimental effect upon the individual stock price. (For example: Exxon Valdez — Exxon Oil.) Companies are now looking for results on a quarter by quarter basis rather than a year-by-year basis. Thus, the safety manager may be required to report more often and collect additional data, which may affect the bottom line.

Conversely, when a company is not doing well and shareholders are jumping ship, the dollars available within the safety budgets may be reduced. In an emergency and disaster situation, this is critical. Shareholders hold your stock to make money. Some shareholders hold for the long run, but more recently, shareholders hold for a short period and trade. A major emergency and disaster situation can immediately place your company stock in play with shareholders making instantaneous decisions to hold or sell your stock. If the situation appears to be appropriately managed, there is a higher likelihood the shareholder will weather the storm. If the shareholder perceives the situation is being managed poorly, the shareholder will sell.

In essence, the shareholders have become your boss. Shareholders normally vote with their feet; thus, if your company incurs a major accident, misses a dividend, or doesn't meet Wall Street's expectations, there is a substantial likelihood that they will sell your stock. When a number of shareholders sell your stock, the value of your company tends to deteriorate, which can influence your borrowing power and your overall operations.

When operations are detrimentally affected, it will have an effect on your budget and thus your overall effectiveness in the safety function.

The key in an emergency situation is to control the flow and type of information being disseminated to the general public and to establish a plan of action to provide your shareholders information regarding the management of the emergency. In essence, this area must be effectively managed just as with all other phases of your emergency and disaster plan.

As with your entire emergency and disaster program, pre-planning is key to effectively minimizing the damage in this area. Below are several items to consider within your program:

- Who will act as the spokesperson for your operation?
- How will the media be controlled on-site?
- How will information be provided to the media?
- What dress will the spokesperson be wearing?
- What background will be utilized for videotape and photographs?
- What brokerage house brought your stock public?
- What brokerage house holds most of your shareholders?
- What lines of communications have been established with the brokerage houses?
- Does your company possess an internet site?
- What information will be provided on the internet?
- How quickly can information be provided to your shareholders?
- Is the information truthful and complete?
- What type of follow-up should be provided to shareholders?
- Is the follow-up timely?
- Is the information being provided to your shareholders reflecting a confident posture?

In today's market, it is much easier for a shareholder to his stock in your company than to risk a loss due to a disaster situation. It is vitally important that careful thought and preparation be provided beforehand to the image the company is sending to the world in 30 second sound bites on television and through the internet. Your shareholders, just like the general public, are making a value judgment about your organization and company from a bit of information. Appropriate management, preparation of image, and effective communication can minimize the overall and long-term damage to your organization.

chapter sixteen

After a disaster — minimizing the damage

> "Look well into thyself; there is a source of strength which will always spring up if thou wilt always look there."
>
> Marcus Aurelius Antoninus

> "The dumbest people I know are those who know it all."
>
> Malcolm Forbes

After the dust of the disaster situation has settled, proper preparation and planning can substantially reduce the cost of the many items, ranging from dealing with employee psychological problems to rebuilding the facility, that are necessary during the salvage phase of the disaster situation. In short, you always pay more and get less if you have to go to the market to acquire goods and services in a short period of time and with the service provider knowing you absolutely need the service. This is very similar to mail service. If you have the time, you can send a letter via U.S. mail for $0.33. If the letter needs to arrive overnight, you will pay a premium price of several dollars to Federal Express, Airborne Express, or another overnight carrier: the same is true in a disaster situation. If the emergency and disaster preparedness plan has addressed the various options and preparations beforehand, a better price can be acquired and the services acquired can be appropriately screened without the stress and urgency of the disaster situation often clouding the decision-making process.

Although every disaster situation is unique, pre-planning and preparation can anticipate many of the issues that may arise during the salvage phase of the operations. Below are several for your review and consideration.

Critical stress debriefing — Your employees and management team members have recently experienced a very stressful situation which may lead to psychological difficulties following the incident. Most psychological illnesses resulting from the disaster situation will be compensated under workers' compensation. Some of the issues to consider include the following:

- Who in your community is qualified to conduct critical stress debriefing?
- Where could critical stress debriefing be conducted?
- Are psychologists or psychiatrists available?
- Where can employees be sent for additional treatment, if necessary?

As part of the overall emergency and disaster preparedness plan, prudent safety professionals can consider entering into a contractual relationship with local psychologists and psychiatrists and other individuals qualified in critical stress debriefing to conduct this necessary service on 24-hour notice. Additionally, prudent safety professionals may consider a short-term lease with a local community center or church to conduct a large scale debriefing session. Acquisition of professionals and locations on short notice often creates difficulties as well as inflated prices. Planning beforehand and entering into appropriate agreements can reduce the cost and ensure appropriate services for your employees and management team members.

Insurance companies — As discussed previously, prudent safety professionals should be acquainted with the extent of coverage and the various provisions of your organization's insurance protections. Pre-planning can assist notifying the appropriate insurance carrier in order to acquire timely benefits and coverage as well as initiate appropriate investigatory activities. Additionally, some provisions of insurance policies provide specific coverage for such issues as lost business, payroll, and other areas, which should be addressed to ensure timely benefits not only to the organization but also to employees and management team members.

Construction equipment — Where can you find a crane when you need one? A bulldozer? During the salvage phase, there is often a need to remove standing structures that are a danger or remove debris to initiate new construction. Who is knowledgeable to operate this equipment even if you are able to locate it?

Prudent safety professionals may want to include a list of local vendors offering heavy equipment for lease along with a listing of local operators. Additionally, a listing of local construction companies with contact names and numbers may also be helpful.

Sale of debris — The old steel beams and copper wiring in a facility that has incurred damage may possess substantial value. Prudent safety professionals may want to include a list of salvage companies for appropriate assessment of the value of the debris and damaged items.

Chapter sixteen: After a disaster — minimizing the damage 93

Information to employees and families — Your employees and their families have just experienced a disaster situation. Many may be injured and some of their friends and co-workers killed. After the initial shock has worn off, reality will set in and they will need information such as where is their pay check?; should they look for a new job?; when can they get their belongings from their locker?

Prudent safety professionals should pre-plan a method to answer all employee questions and provide pertinent information to employees after a disaster situation. If employees are trained in communicating as part of their emergency and disaster program training, they will remember the communications method. For example, all employees will receive a telephone call within 48 hours of the disaster; all employees will receive information on meeting times on the local radio station. If the method of communications is pre-planned, employees will be relieved from the stress of not knowing what to do after the disaster situation.

To establish an *after disaster* communication system requires a substantial amount of forethought and planning on the part of the safety professional. For example, if the emergency and disaster plan established a telephone notification within 48 hours, who is responsible for making the telephone calls? Who is going to acquire the information and train the callers? Where are the telephone calls going to be made from (i.e., location)?

The worst possible scenario is for employees to experience a disaster situation, experience the stress following the disaster, and sit at home not knowing where their next pay check will be coming from and if they still have a job. Appropriate and timely communication following a disaster situation is crucial and should not be taken lightly in the pre-planning stage of the emergency and disaster program.

- Payroll — All of your records are in the facility that has been destroyed. Your back-up records are in corporate headquarters across the country. Employees are scheduled to be paid tomorrow. Prudent safety professionals may wish to plan for alternative methods of acquiring and providing payroll to their employees including back-up services, alternate locations of distribution, and related services.
- Informing customers — After a disaster situation, larger companies are often able to shift their distribution of products to other facilities. Smaller companies do not possess this luxury and will need to contact their customers and distributors to provide information regarding the future production and shipment of their products.
- Airlift/transport of injured employees — In smaller communities, local hospital facilities may not be capable of providing appropriate care for specific injuries or critically injured employees. These injured employees will require transport to regional or national hospitals for care. Prudent safety professionals may want to include a listing of the various air transport and specialty transport companies in their pre-planning documents.

Appropriate planning following a disaster situation is important in order to minimize the cost and provide all of the necessary services for your employees and management team members. Planning for after a disaster should be given the same effort as other parts of the emergency and disaster preparedness planning. Remember, the disaster situation does not end when the fire trucks roll up their hoses and go home.

chapter seventeen

Governmental reactions

> "The real problem with which modern government has to deal is how to protect the citizen against the encroachment upon his rights and liberties by his own government, how to save him from the repressive schemes born of the egotism of public office."
>
> Eugene R. Black

> "If we fixed a hangnail the way our government fixes the economy, we'd slam a car door on it."
>
> Cullen Hightower

> "I think we have more machinery of government than is necessary, too many parasites living on the labors of the industrious."
>
> Thomas Jefferson

When a disaster occurs, *everyone who is anyone* will be responding to the scene. The initial response is usually the fire, EMS, and law enforcement to react to the injuries, fire, explosions, or other sources. As discussed above, these vital public sector organizations become part and parcel of your overall response plan. The *second wave* of response to a disaster situation can include a myriad of governmental agencies ranging from OSHA (Occupational Safety and Health Administration) to local prosecutors whose tasks usually involve investigative and exploratory after-the-disaster duties. For many organizations, the results of this investigative phase can be as damaging as the disaster itself if the deluge of inquiry has not been adequately anticipated.

Safety professionals should be aware of their rights and responsibilities when addressing OSHA or other governmental agencies before and after a disaster situation. It is important that the safety professional be aware of the underlying purpose of the investigation and the potential civil and criminal penalties which can be assessed against the organization.

The Occupational Safety and Health Act of 1970 (OSHA Act) covers virtually every American workplace that employs one or more employees and engages in a business that in any way affects interstate commerce.[1] The OSHA Act covers employment in every state, the District of Columbia, Puerto Rico, Guam, the Virgin Islands, American Samoa, and the Trust Territory of the Pacific Islands.[2] The OSHA Act does not, however, cover employees in situations where other state or federal agencies have jurisdiction.[3] Additionally, the OSHA Act exempts residential owners who employ people for ordinary domestic tasks, such as cooking, cleaning, and child care.[4] It also does not cover federal,[5] state, and local governments[6] or Native American reservations.[7]

The OSHA Act does require every employer engaged in interstate commerce to furnish employees "a place of employment ... free from recognized hazards that are causing, or are likely to cause, death or serious harm."[8] To help employers create and maintain safe working environments and to enforce laws and regulations that ensure safe and healthful work environments, Congress provided for the creation of OSHA, to be a new agency under the direction of the Department of Labor.

Today, OSHA is one of the most widely known and powerful enforcement agencies. It has been granted broad regulatory powers to promulgate regulations and standards, investigate and inspect, issue citations, and propose penalties for safety violations in the workplace. The OSHA Act also established an independent agency to review OSHA citations and decisions, the Occupational Safety and Health Review Commission (OSHRC). The OSHRC is a quasi-judicial and independent administrative agency composed of three commissioners appointed by the President who serve staggered six year terms. The OSHRC has the power to issue orders, uphold, vacate, or modify OSHA citations and penalties, and direct other appropriate relief and penalties.

The educational arm of the OSHA Act is the National Institute for Occupational Safety and Health (NIOSH), which was created as a specialized educational agency of the existing National Institute of Health. NIOSH conducts occupational safety and health research and develops criteria for new OSHA standards. NIOSH can conduct workplace inspections, issue subpoenas,

---

[1] 29 C.F.R. 1910.
[2] Ibid. Section 652(7).
[3] *See e.g.,* Atomic Energy Act of 1954, 42 U.S.C. Section 2021.
[4] 29 C.F.R. Section 1975(6).
[5] 29 U.S.C.A. Section 652(5) (no coverage under OSHA Act, when U.S. government acts as employer.)
[6] Ibid.
[7] *See e.g., Navajo Forest Prods. Indus.*, 8 OSH Cases 2694 (OSH Rev. Comm'n 1980), aff'd, 692 F.2d 709, 10 OSH Cases 2159.
[8] 29 U.S.C.A. Section 654(a)(1).

Chapter seventeen: Governmental reactions 97

and question employees and employers, but it does not have the power to issue citations or penalties.

As permitted under the OSHA Act, OSHA encourages individual states to take responsibility for OSHA administration and enforcement within their own respective boundaries. Each state possesses the ability to request and be granted the right to adopt state safety and health regulations and enforcement mechanisms.[1] For a state plan to be placed into effect, the state must first develop and submit its proposed program to the Secretary of Labor for review and approval. The Secretary must certify that the state plan's standards are at least as effective as the federal standards and that the state will devote adequate resources to administering and enforcing standards.[2]

In most state plans, the state agency has developed more stringent safety and health standards than OSHA[3] and has usually developed more stringent enforcement schemes.[4] The Secretary of Labor has no statutory authority to reject a state plan if the proposed standards or enforcement schemes are stricter than the OSHA standards, but can reject the state plan if the standards are below the minimum limits set under OSHA standards.[5] These states are known as "state plan" states and territories.[6] (As of 1991, there were 21 states and two territories with approved and functional state plan programs.[7])

---

[1] In section 18(b), the OSHA Act provides that any state "which, at any time, desires to assume responsibility for development and the enforcement therein of occupational safety and health standards relating to any... issue with respect to which a federal standard has been promulgated... shall submit a state plan for the development of such standards and their enforcement.")

[2] 8 Ibid. Section 667(c). After an initial evaluation period of at least three years during which OSHA retains concurrent authority, a state with an approved plan gains exclusive authority over standard setting, inspection procedures, and enforcement of health and safety issues covered under the state plan. *See Also Noonan v. Texaco*, 713 P.2d 160 (Wyo. 1986); Plans for the Development and Enforcement of State Standards, 29 C.F.R. Section 667(f) (1982) and section 1902.42(c)(1986). Although the state plan is implemented by the individual state, OSHA continues to monitor the program and may revoke the state authority if the state does not fulfill the conditions and assurances contained within the proposed plan.

[3] Some states incorporate federal OSHA standards into their plans and add only a few of their own standards as a supplement. Other states, such as Michigan and California, have added a substantial number of separate and independently promulgated standards. *See generally* Employee Safety and Health Guide (CCH) sections 5000-5840 (1987) (compiling all state plans). Some states also add their own penalty structures. For example, under Arizona's plan, employers may be fined up to $150,000 and sentenced to one and one-half years in prison for knowing violations of state standards that cause death to an employee and may also have to pay $25,000 in compensation to the victim's family. If the employer is a corporation, the maximum fine is $1 million. *See* Ariz. Rev. Stat. Ann. Sections 13-701, 13-801, 23-4128, 23-418.01, 13-803 (Supp. 1986).

[4] For example, under Kentucky's state plan regulations for controlling hazardous energy (i.e., lockout/tagout), locks would be required rather than locks or tags being optional as under the federal standard. Lockout/tagout is discussed in more detail in Chapter 2.

[5] 29 U.S.C. Section 667.

[6] 29 U.S.C.A. Section 667; 29 C.F.R. Section 1902.

[7] The states and territories operating their own OSHA programs are Alaska, Arizona, California, Hawaii, Indiana, Iowa, Kentucky, Maryland, Michigan, Minnesota, Nevada, New Mexico, North Carolina partial federal OSHA enforcement), Oregon, Puerto Rico, South Carolina, Tennessee, Utah, Vermont, Virginia, Virgin Islands, Washington, and Wyoming.

Employers in state plan states and territories must comply with their state's regulations; federal OSHA plays virtually no role in direct enforcement.

OSHA does, however, possess an approval and oversight role with regards to state plan programs. OSHA must approve all state plan proposals prior to enactment. OSHA maintains oversight authority to *pull the ticket* of any/all state plan programs at any time if they are not meeting the required prerequisites. Enforcement of this oversight authority was recently observed following the fire resulting in several workplace fatalities at the Imperial Foods facility in Hamlet, NC. Following this incident, OSHA assumed jurisdiction and control over the state plan program in North Carolina and made significant modifications to this program before returning the program to state control.

The OSHA Act requires that a covered employer comply with specific occupational safety and health standards and all rules, regulations, and orders issued pursuant to the OSHA Act that apply to the workplace.[1] The OSHA Act also requires that all standards be based on research, demonstration, experimentation, or other appropriate information.[2] The Secretary of Labor is authorized under the Act to "promulgate, modify, or revoke any occupational safety and health standard,"[3] and the OSHA Act describes the procedures that the Secretary must follow when establishing new occupational safety and health standards.[4]

The OSHA Act authorizes three ways to promulgate new standards. From 1970 to 1973, the Secretary of Labor was authorized in section 6(a) of the Act[5] to adopt national consensus standards and establish federal safety and health standards without following lengthy rulemaking procedures. Many of the early OSHA standards were adapted mainly from other areas of regulation, such as the National Electric Code and American National Standards Institute (ANSI) guidelines. However, this promulgation method is no longer in effect.

The usual method of issuing, modifying, or revoking a new or existing OSHA standard is set out in section 6(b) of the OSHA Act and is known as informal rulemaking. It requires notice to interested parties, through subscription in the *Federal Register* of the proposed regulation and standard, and provides an opportunity for comment in a non-adversarial administrative hearing.[6] The proposed standard can also be advertised through magazine articles and other publications, thus informing interested parties of the proposed standard and regulation. This method differs from the requirements of most other administrative agencies that follow the Administrative Procedure

---

[1] 29 U.S.C. Section 655(b).
[2] 29 U.S.C.A. Section 655(b)(5).
[3] 29 U.S.C. 1910.
[4] 29 C.F.R. Section 1911.15. (By regulation, the Secretary of Labor has prescribed more detailed procedures than the OSHA Act specifies to ensure participation in the process of setting new standards, 29 C.F.R. Section 1911.15.)
[5] 29 U.S.C. Section 1910.
[6] 29 U.S.C. Section 655(b).

Act[1] in that the OSHA Act provides interested persons an opportunity to request a public hearing with oral testimony. It also requires the Secretary of Labor to publish in the *Federal Register* a notice of the time and place of such hearings.

Although not required under the OSHA Act, the Secretary of Labor has directed, by regulation, that OSHA follow a more rigorous procedure for comment and hearing than other administrative agencies.[2] Upon notice and request for a hearing, OSHA must provide a hearing examiner in order to listen to any oral testimony offered. All oral testimony is preserved in a verbatim transcript. Interested persons are provided an opportunity to cross-examine OSHA representatives or others on critical issues. The Secretary must state the reasons for the action to be taken on the proposed standard, and the statement must be supported by substantial evidence in the record as a whole.

The Secretary of Labor has the authority not to permit oral hearings and to call for written comment only. Within 60 days after the period for written comment or oral hearings has expired, the Secretary must decide whether to adopt, modify, or revoke the standard in question. The Secretary can also decide not to adopt a new standard. The Secretary must then publish a statement of the reasons for any decision in the *Federal Register*. OSHA regulations further mandate that the Secretary provide a supplemental statement of significant issues in the decision. Safety and health professionals should be aware that the standard as adopted and published in the *Federal Register* may be different from the proposed standard. The Secretary is not required to reopen hearings when the adopted standard is a *logical outgrowth* of the proposed standard.[3]

The final method for promulgating new standards, and the one most infrequently used, is the emergency temporary standard permitted under section 6(c).[4] The Secretary of Labor may establish a standard immediately if it is determined that employees are subject to grave danger from exposure to substances or agents known to be toxic or physically harmful and that an emergency standard would protect the employees from the danger. An emergency temporary standard becomes effective on publication in the *Federal Register* and may remain in effect for six months. During this six-month period, the Secretary must adopt a new permanent standard or abandon the emergency standard.

Only the Secretary of Labor can establish new OSHA standards. Recommendations or requests for an OSHA standard can come from any interested person or organization, including employees, employers, labor unions, environmental groups, and others.[5] When the Secretary receives a petition to adopt a new standard or to modify or revoke an existing standard, he or she

[1] 5 U.S.C. Section 553.
[2] 29 C.F.R. Section 1911.15.
[3] *Taylor Diving & Salvage Co. v. Department of Labor*, 599 F.2d 622 7 OSH Cases 1507 (5th Cir. 1979).
[4] 29 U.S.C. Section 655(c).
[5] Ibid. at section 655(b)(1).

Table 17.1 Violation and Penalty Schedule

Penalty	Old Penalty Schedule (in dollars)	New Penalty Schedule (1990) (in dollars)
De minimis notice	0	0
Non-serious	0–1,000	0–7,000
Serious	0–1,000	0–7,000
Repeat	0–10,000	0–70,000
Willful	0–10,000	25,000 minimum 70,000 maximum
Failure to abate notice	0–1,000 per day	0–7,000 per day
New posting penalty		0–7,000

usually forwards the request to NIOSH and the National Advisory Committee on Occupational Safety and Health (NACOSH)[1] or the Secretary may use a private organization such as American National Standards Institute (ANSI) for advice and review.

The OSHA Act requires that an employer must maintain a place of employment free from recognized hazards that are causing or are likely to cause death or serious physical harm, even if there is no specific OSHA standard addressing the circumstances. Under section 5(a)(1), known as the *general duty clause*, an employer may be cited for a violation of the OSHA Act if the condition causes harm or is likely to cause harm to employees, even if OSHA has not promulgated a standard specifically addressing the particular hazard. The general duty clause is a catch-all standard encompassing all potential hazards that have not been specifically addressed in the OSHA standards. For example, if a company is cited for an ergonomic hazard and there is no ergonomic standard to apply, the hazard will be cited under the general duty clause.

The OSHA Act provides for a wide range of penalties, from a simple notice with no fine to criminal prosecution. The Omnibus Budget Reconciliation Act of 1990 multiplied maximum penalties sevenfold. This law summarily modified the entire OSHA penalty schedule. Violations are categorized and penalties may be assessed as outlined in Table 17.1.

Each alleged violation is categorized and the appropriate fine issued by the OSHA area director. It should be noted that each citation is separate and may carry with it a monetary fine. The gravity of the violation is the primary factor in determining penalties.[2] In assessing the gravity of a violation, the compliance officer or area director must consider (1) the severity of the injury or illness that could result and (2) the probability that an injury or illness could occur as a result of the violation.[3] Specific penalty assessment tables

---

[1] Ibid. at section 656(a)(1). NACOSH was created by the OSHA Act to "advise, consult with, and make recommendations...on matters relating to the administration of the Act." Normally, for new standards, the Secretary has established continuing committees and ad hoc committees to provide advice regarding particular problems or proposed standards.
[2] *OSHA Compliance Field Operations Manual (OSHA Manual)* at XI-C3c (April 1977).
[3] Ibid.

Chapter seventeen: Governmental reactions

assist the area director or compliance officer in determining the appropriate fine for the violation.[1]

After selecting the appropriate penalty table, the area director or other official determines the degree of probability that the injury or illness will occur by considering:

1. The number of employees exposed
2. The frequency and duration of the exposure
3. The proximity of employees to the point of danger
4. Factors such as the speed of the operation that require work under stress
5. Other factors that might significantly affect the degree of probability of an accident[2]

OSHA has defined a serious violation as "an infraction in which there is a substantial probability that death or serious harm could result ... unless the employer did not or could not with the exercise of reasonable diligence, know of the presence of the violation."[3] section 17(b) of the OSHA Act requires that a penalty of up to $7000 be assessed for every serious violation cited by the compliance officer.[4] In assembly line enterprises and manufacturing facilities with duplicate operations, if one process is cited as possessing a serious violation, it is possible that each of the duplicate processes or machines may be cited for the same violation. Thus, if a serious violation is found in one machine and there are many other identical machines in the enterprise, a very large monetary fine for a single serious violation is possible.[5]

Currently the greatest monetary liabilities are for *repeat violations, willful violations,* and *failure to abate* cited violations. A **repeat** violation is a second citation for a violation that was cited previously by a compliance officer. OSHA maintains records of all violations and must check for repeat violations after each inspection. A **willful** violation is the employer's purposeful or negligent failure to correct a known deficiency. This type of violation, in addition to carrying a large monetary fine, exposes the employer to a charge of an *egregious* violation and the potential for criminal sanctions under the OSHA Act or state criminal statutes if an employee is injured or killed as a direct result of the willful violation. **Failure to abate** a cited violation has the greatest cumulative monetary liability of all. OSHA may assess a penalty of up to $1000 per day per violation for each day in which a cited violation is not brought into compliance.

---

[1] Ibid. at XI-C3c(2).
[2] Ibid. at (3)(a).
[3] 29 U.S.C. Section 666.
[4] Ibid. at section 666(b).
[5] For example, if a company possesses 25 identical machines, and each of these machines is found to have the identical serious violation, this would theoretically constitute 25 violations rather than one violation on 25 machines, and a possible monetary fine of $175,000 rather than a maximum of $7000.

In assessing monetary penalties, the area or regional director must consider the good faith of the employer, the gravity of the violation, the employer's past history of compliance, and the size of the employer. Mr. Joseph Dear, the former Assistant Secretary of Labor, recently stated that OSHA will start using its egregious case policy, which has seldom been invoked in recent years.[1] Under the egregious violation policy, when violations are determined to be conspicuous, penalties are cited for each violation, rather than combining the violations into a single, smaller penalty.

In addition to the potential civil or monetary penalties that could be assessed, OSHA regulations may be used as evidence in negligence, product liability, workers' compensation, and other actions involving employee safety and health issues.[2] OSHA standards and regulations are the baseline requirements for safety and health that must be met, not only to achieve compliance with the OSHA regulations, but also to safeguard an organization against other potential civil actions.

The OSHA monetary penalty structure is classified according to the type and gravity of the particular violation. Violations of OSHA standards or the general duty clause are categorized as de minimis,[3] other (non-serious),[4] serious,[5] repeat,[6] and willful.[7] See Tables 17.1 for penalty schedules and tables. Monetary penalties assessed by the Secretary vary according to the degree of the violation. Penalties range from no monetary penalty to 10 times the imposed penalty for repeat or willful violations.[8] Additionally, the Secretary may refer willful violations to the U.S. Department of Justice for imposition of criminal sanctions.[9]

De minimis violations

When a violation of an OSHA standard does not immediately or directly relate to safety or health, OSHA either does not issue a citation or issues a de minimis citation. Section 9 of the OSHA Act provides that "[the] Secretary may prescribe procedures for the issuance of a notice in lieu of a citation with respect to de minimis violations which have no direct or immediate relationship to safety or health."[10] A de minimis notice does not constitute a citation and no fine is imposed. Additionally, there usually is no abatement period and thus there can be no violation for failure to abate.

---

[1] *Occupational Safety & Health Reporter*, V. 23, No. 32, Jan. 12, 1994.
[2] See Infra at section 1.140.
[3] 29 U.S.C. Sections 658(a), 666(c).
[4] Ibid. at section 666(j).
[5] Ibid. at section 666(c).
[6] Ibid. at (a).
[7] Ibid.
[8] Ibid. at (b).
[9] Ibid. at (e).
[10] Ibid. at section 658(a).

The *OSHA Compliance Field Operations Manual* (*OSHA Manual*)[1] provides two examples of when de minimis notices are generally appropriate:

1. "in situations involving standards containing physical specificity wherein a slight deviation would not have an immediate or direct relationship to safety or health,"[2] and
2. "where the height of letters on an exit sign is not in strict conformity with the size requirements of the standard."[3]

OSHA has found de minimis violations in cases where employees, as well as the safety records, are persuasive in exemplifying that no injuries or lost time have been incurred.[4] Additionally, in order for OSHA to conserve valuable resources to produce a greater impact on safety and health in the workplace, it is highly likely that the Secretary will encourage use of the de minimis notice in marginal cases and even in other situations where the possibility of injury is remote and potential injuries would be minor.

Other or non-serious violations

Other or non-serious violations are issued when a violation could lead to an accident or occupational illness, but the probability that it would cause death or serious physical harm is minimal. Such a violation, however, does possess a direct or immediate relationship to the safety and health of workers.[5] Potential penalties for this type of violation range from no fine up to $7000 per violation.[6]

In distinguishing between a serious and a non-serious violation, the OSHRC has stated that "a non-serious violation is one in which there is a direct and immediate relationship between the violative condition and occupational safety and health but no such relationship that a resultant injury or illness is death or serious physical harm."[7]

The *OSHA Manual* provides guidance and examples for issuing non-serious violations. It states:

> an example of non-serious violation is the lack of guardrail at a height from which a fall would more probably result in only a mild sprain or cut or abrasion; i.e., something less than serious harm.[8]

---

[1] *Supra* note 62.
[2] Ibid. at VII-B3a.
[3] Ibid.
[4] *Hood Sailmakers*, 6 OSH Cases 1207 (1977).
[5] *OSHA Manual supra* note 62, at VIII-B2a. The proper nomenclature for this type of violation is *other* or *other than serious*. Many safety and health professionals classify this type of violation as nonserious for explanation and clarification purposes.
[6] A nonserious penalty is usually less than $100 per violation.
[7] *Crescent Wharf & Warehouse Co.*, 1 OSH Cases 1219, 1222 (1973).
[8] *OSHA Manual, supra* at n. 62, at VIII-B2a.

> A citation for serious violation may be issued or a group of individual violations (which) taken by themselves would be non-serious, but together would be serious in the sense that in combination they present a substantial probability of injury resulting in death or serious physical harm to employees.[1]
>
> A number of non-serious violations (which) are present in the same piece of equipment which, considered in relation to each other, affect the overall gravity of possible injury resulting from an accident involving the combined violations...may be grouped in a manner similar to that indicated in the preceding paragraph, although the resulting citation will be for a non-serious violation.[2]

The difference between a serious and a non-serious violation hinges on subjectively determining the probability of injury or illness that might result from the violation. Administrative decisions have usually turned on the particular facts of the situation. OSHRC has reduced serious citations to non-serious violations when the employer was able to show that the probability of an accident, and the probability of a serious injury or death, was minimal.[3]

Serious violations

Section 17(k) of the OSHA Act defines a serious violation as one where:

> there is a substantial probability that death or serious physical harm could result from a condition which exists, or from one or more practices, means, methods, operations or processes which have been adopted or are in use, in such place of employment unless the employer did not, and could not with exercise of reasonable diligence, know of the presence of the violation.[4]

Section 17(b) of the Act provides that a civil penalty of up to $7000 must be assessed for serious violations, whereas for non-serious violations civil penalties may be assessed.[5] The amount of the penalty is determined by considering (1) the gravity of the violation, (2) the size of the employer, (3) the good faith of the employer, and (4) the employer's history of previous violations.[6]

[1] Ibid. at B2b(1).
[2] Ibid. at (2).
[3] See *Secretary v. Diamond In.*, 4 OSH Cases 1821 (1976); *Secretary v. Northwest Paving*, 2 OSH Cases 3241 (1974); *Secretary v. Sky-Hy Erectors & Equip.*, 4 OSH Cases 1442 (1976). But see *Shaw Constr. v. OSHRC*, 534 F.2d 1183, 4 OSH Cases 1427 (5th Cir. 1976)(holding that serious citation was proper whenever accident was merely possible.)
[4] 29 U.S.C. Section 666(j).
[5] Ibid.
[6] Ibid. at section 666(i).

To prove that a violation is within the serious category, OSHA must only show a substantial probability that a foreseeable accident would result in serious physical harm or death. Thus, contrary to common belief, OSHA does not need to show that a violation would create a high probability that an accident would result. Because substantial physical harm is the distinguishing factor between a serious and a non-serious violation, OSHA has defined *serious physical harm* as "permanent, prolonged, or temporary impairment of the body in which part of the body is made functionally useless or is substantially reduced in efficiency on or off the job." Additionally, an occupational illness is defined as "illness that could shorten life or significantly reduce physical or mental efficiency by inhibiting the normal function of a part of the body."[1]

After determining that a hazardous condition exists and that employees are exposed or potentially exposed to the hazard, the *OSHA Manual* instructs compliance officers to use a four-step approach to determine whether the violation is serious:

1. Determine the type of accident or health hazard exposure that the violated standard is designed to prevent in relation to the hazardous condition identified
2. Determine the type of injury or illness which is reasonably predictable and could result from the type of accident or health hazard exposure identified in step 1
3. Determine if the type of injury or illness identified in step 2 includes death or a form of serious physical harm
4. Determine if the employer knew or with the exercise of reasonable diligence could have known of the presence of the hazardous condition[2]

The *OSHA Manual* provides examples of serious injuries, including amputations, fractures, deep cuts involving extensive suturing, disabling burns, and concussions. Examples of serious illnesses include cancer, silicosis, asbestosis, poisoning, and hearing and visual impairment.[3]

Safety professionals should be aware that OSHA is not required to show that the employer actually knew that the cited condition violated safety or health standards. The employer can be charged with constructive knowledge of the OSHA standards. OSHA also does not have to show that the employer could reasonably foresee that an accident would happen, although it does have the burden of proving that the possibility of an accident was not totally

---

[1] *OSHA Manual, supra* at n. 62, at IV-B-1(b)(3)(a),(c).
[2] Ibid. at VIII-B1b(2)(c). In determining whether a violation constitutes a serious violation, the compliance officer is functionally describing the prima facie case that the Secretary would be required to prove, i.e., (1) the casual link between the violation of the safety or health standard and the hazard, (2) reasonably predictable injury or illness that could result, (3) potential of serious physical harm or death, and (4) the employer's ability to foresee such harm by using reasonable diligence.
[3] Ibid. at VIII-B1c(3)a.

unforeseeable. OSHA does need to prove, however, that the employer knew or should have known of the hazardous condition and that it knew there was a substantial likelihood that serious harm or death would result from an accident.[1] If the Secretary cannot prove that the cited violation meets the criteria for a serious violation, the violation may be cited in one of the lesser categories.

Willful violations

The most severe monetary penalties under the OSHA penalty structure are for willful violations. A *willful* violation can result in penalties of up to $70,000 per violation, with a minimum required penalty of $5000. Although the term willful is not defined in OSHA regulations, courts generally have defined a willful violation as "an act voluntarily with either an intentional disregard of, or plain indifference to, the Act's requirements."[2] Further, the OSHRC defines a willful violation as "action taken knowledgeably by one subject to the statutory provisions of the OSHA Act in disregard of the action's legality. No showing of malicious intent is necessary. A conscious, intentional, deliberate, voluntary decision is properly described as willful."[3]

There is little distinction between civil and criminal willful violations other than the due process requirements for a criminal violation and the fact that a violation of the general duty clause cannot be used as the basis for a criminal willful violation. The distinction is usually based on the factual circumstances and the fact that a criminal willful violation results from a willful violation which caused an employee death.

According to the *OSHA Manual*, the compliance officer "can assume that an employer has knowledge of any OSHA violation condition of which its supervisor has knowledge; he can also presume that, if the compliance officer was able to discover a violative condition, the employer could have discovered the same condition through the exercise of reasonable diligence."[4]

Courts and the OSHRC have agreed on three basic elements of proof that OSHA must show for a willful violation. OSHA must show that the employer (1) knew or should have known that a violation existed, (2) voluntarily chose not to comply with the OSHA Act to remove the violative condition, and (3) made the choice not to comply with intentional disregard of the OSHA Act's requirements or plain indifference to them properly characterized as reckless. Courts and the OSHRC have affirmed findings of willful violations in many circumstances, ranging from deliberate disregard

---

[1] Ibid. at (4). *See also, Cam Indus.*, 1 OSH Cases 1564 (1974); *Secretary v. Sun Outdoor Advertising*, 5 OSH Cases 1159 (1977).
[2] *Cedar Constr. Co. v. OSHRC*, 587 F.2d 1303, 6 OSH Cases 2010, 2011 (D.C. Cir. 1971). Moral turpitude or malicious intent are not necessary elements for a willful violation. *U.S. v. Dye Constr.*, 522 F.2d 777, 3 OSH Cases 1337 (4th Cir. 1975); *Empire-Detroit Steel v. OSHRC*, 579 F.2d 378, 6 OSH Cases 1693 (6th Cir. 1978).
[3] *P.A.F. Equip. Co.*, 7 OSH Cases 1209 (1979).
[4] *OSHA Manual*, supra at n. 62, at VIII-B1c(4).

of known safety requirements[1] through fall protection equipment not being provided.[2] Other examples of willful violations include cases where safety equipment was ordered but employees were permitted to continue work until the equipment arrived,[3] inexperienced and untrained employees were permitted to perform a hazardous job,[4] and where an employer failed to correct a situation that had been previously cited as a violation.

Repeat and failure to abate violations

Repeat and *failure to abate* violations are often quite similar and confusing to human resource (HR) professionals. When, upon reinspection by OSHA, a violation of a previously cited standard is found but the violation does not involve the same machinery, equipment, process, or location, it would constitute a repeat violation. If, upon reinspection by OSHA, a violation of a previously cited standard is found but evidence indicates that the violation continued uncorrected since the original inspection, it would constitute a failure to abate violation.[5]

The most costly civil penalty under the OSHA Act is for repeat violations. The OSHA Act authorizes a penalty of up to $70,000 per violation but permits a maximum penalty of 10 times the maximum authorized for the first instance of the violation. Repeat violations can also be grouped within the willful category (i.e., a willful repeat violation) to acquire maximum civil penalties.

In certain cases where an employer has more than one fixed establishment and citations have been issued, the *OSHA Manual* states,

> for the purpose for considering whether a violation is repeated, citations issued to employers having fixed establishments (e.g., factories, terminals, stores) will be limited to the cited establishment. For employers engaged in businesses having no fixed establishments, repeated violations will be alleged based upon prior violations occurring anywhere within the same Area Office Jurisdiction.[6]

When a previous citation has been contested but a final OSHRC order has not yet been received, a second violation is usually cited as a repeat violation. The *OSHA Manual* instructs the compliance officer to notify the assistant regional director and to indicate in the citation that the violation is contested.[7] If the first citation never becomes a final OSHRC order (i.e., the citation is

---

[1] *Universal Auto Radiator Mfg. Co. v. Marshall*, 631 F.2d 20, 8 OSH Cases 2026 (3d Cir. 1980).
[2] *Haven Steel Co. v. OSHRC*, 738 F.2d 397, 11 OSH Cases 2057 (10th Cir. 1984).
[3] *Donovan v. Capital City Excavating Co.*, 712 F.2d 1008, 11 OSH Cases 1581 (6th Cir. 1983).
[4] *Ensign-Bickford Co. v. OSHRC*, 717 F.2d 1419, 11 OSH Cases 1657 (D.C. Cir. 1983).
[5] *OSHA Manual, supra* at n. 62, at VIII-B5c.
[6] Ibid. at IV-B5(c)(1).
[7] Ibid. at VIII-B5d.

vacated or otherwise dismissed), the second citation for the repeat violation will be removed automatically.[1]

As noted previously, a failure to abate violation occurs when, upon reinspection, the compliance officer finds that the employer has failed to take necessary corrective action and thus the violation continues uncorrected. The penalty for a failure to abate violation can be up to $7000 per day to a maximum of $70,000. HR professionals should also be aware that citations for repeat violations, failure to abate violations, or willful repeat violations can be issued for violations of the general duty clause. The *OSHA Manual* instructs compliance officers that citations under the general duty clause are restricted to serious violations or to willful or repeat violations that are of a serious nature.[2]

Failure to post violation notices

A new penalty category, the failure to post violation notices carries a penalty of up to $7000 for each violation. A failure to post violation occurs when an employer fails to post notices required by the OSHA standards, including the OSHA poster, a copy of the year end summary of the OSHA 200 form, a copy of OSHA citations when received, and copies of other pleadings and notices.

Criminal liability and penalties

The OSHA Act provides for criminal penalties in four circumstances.[3] In the first, anyone inside or outside of the Department of Labor or OSHA who gives advance notice of an inspection, without authority from the Secretary, may be fined up to $1000, imprisoned for up to six months, or both. Second, where any employer or person, who intentionally falsifies statements or OSHA records that must be prepared, maintained, or submitted under the OSHA Act, may if found guilty be fined up to $10,000, imprisoned for up to six months, or both. Third, when any person responsible for a violation of an OSHA standard, rule, order, or regulation, causes the death of an employee may, upon conviction, be fined up to $10,000, imprisoned for up to six months, or both. If convicted for a second violation, punishment may be a fine of up to $20,000, imprisonment for up to one year, or both.[4] Finally, if an individual is convicted of forcibly resisting or assaulting a compliance officer or other Department of Labor personnel, a fine of $5000, three years in prison, or both can be imposed. Any person convicted of killing a compliance officer or other OSHA or Department of Labor personnel acting in his or her official capacity may be sentenced to prison for any term of years or life.

[1] Ibid.
[2] Ibid. at XI-C5c.
[3] 29 U.S.C. Section 666(e)--(g). *See also, OSHA Manual* ,supra note 62 at VI-B.
[4] A repeat criminal conviction for a willful violation causing an employee death doubles the possible criminal penalties.

Chapter seventeen: Governmental reactions

OSHA does not have authority to impose criminal penalties directly, instead, it refers cases for possible criminal prosecution to the U.S. Department of Justice. Criminal penalties must be based on violation of a specific OSHA standard; they may not be based on a violation of the general duty clause. Criminal prosecutions are conducted like any other criminal trial, with the same rules of evidence, burden of proof, and rights of the accused. A corporation may be criminally liable for the acts of its agents or employees.[1] The statute of limitations for possible criminal violations of the OSHA Act, as for other federal noncapital crimes, is five years.[2]

Under federal criminal law, criminal charges may range from murder to manslaughter to conspiracy. Several charges may be brought against an employer for various separate violations under one federal indictment.

The OSHA Act provides for criminal penalties of up to $10,000 and/or imprisonment for up to six months. A repeated willful violation causing an employee death can double the criminal sanction to a maximum of $20,000 and/or one year of imprisonment. Given the increased use of criminal sanctions by OSHA in recent years, personnel managers should advise their employers about the potential for these sanctions being used when the safety and health of employees are disregarded or put on the back burner.

Criminal liability for a willful OSHA violation can attach to an individual or a corporation. In addition, corporations may be held criminally liable for the actions of their agents or officials.[3] Safety professionals and other corporate officials may also be subject to criminal liability under a theory of aiding and abetting the criminal violation in their official capacity with the corporation.[4]

Safety professionals should also be aware that an employer could face two prosecutions for the same OSHA violation without the protection of double jeopardy. The OSHRC can bring an action for a civil willful violation using the monetary penalty structure described previously and the case then may be referred to the Justice Department for criminal prosecution of the same violation.[5]

Prosecution of willful criminal violations by the Justice Department has been rare in comparison to the number of inspections performed and violations cited by OSHA on a yearly basis. However, the use of criminal sanctions has increased substantially in the last few years. With adverse publicity being generated as a result of workplace accidents and deaths[6] and Congress emphasizing reform, a decrease in criminal prosecutions is unlikely.

---

[1] 29 C.F.R. Section 5.01(6).
[2] *U.S. v. Dye Const. Co.*, 510 F.2d 78, 2 OSH Cases 1510 (10th Cir. 1975).
[3] *U.S. v. Crosby & Overton*, No. CR-74-1832-F (S.D. Cal. Feb. 24, 1975.)
[4] 18 U.S.C. Section 2.
[5] These are uncharted waters. Employers may argue due process and double jeopardy, but OSHA may argue that it has authority to impose penalties in both contexts. There are currently no cases on this issue.
[6] Jefferson, *Dying for Work*, A.B.A. #J. 46 (Jan. 1993).

The law regarding criminal prosecution of willful OSHA Act violations is still emerging. Although few cases have actually gone to trial, in most situations the mere threat of criminal prosecution has encouraged employers to settle cases with assurances that criminal prosecution would be dismissed. Many state plan states are using criminal sanctions permitted under their state OSH regulations more frequently.[1] State prosecutors have also allowed use of state criminal codes for workplace deaths.[2]

After a disaster situation, especially if a fatality is involved, safety professionals should exercise extreme caution. The potential for criminal sanctions and criminal prosecution is substantial if a willful violation of a specific OSHA standard is directly involved in the death. The OSHA investigation may be conducted from a criminal perspective in order to gather and secure the appropriate evidence to later pursue criminal sanctions.[3] A prudent safety professional facing a workplace fatality investigation should address the OSHA investigation with legal counsel present and reserve all rights guaranteed under the U.S. Constitution.[4] Obviously, under no circumstances should a safety professional condone or attempt to conceal facts or evidence which consists of a cover-up.

OSHA performs all enforcement functions under the OSHA Act. Under section 8(a) of the Act, OSHA compliance officers have the right to enter any workplace of a covered employer without delay, inspect and investigate a workplace during regular hours and at other reasonable times, and obtain an inspection warrant if access to a facility or operation is denied.[5] Upon arrival at an inspection site, the compliance officer is required to present his or her credentials to the owner or designated representative of the employer before starting the inspection. The employer representative and an employee and/or union representative may accompany the compliance officer on the inspection. Compliance officers can question the employer and employees and inspect required records, such as the OSHA Form 200, which records injuries and illnesses.[6] Most compliance officers cannot issue on-the-spot citations, they only have authority to document potential hazards and report or confer with the OSHA area director before issuing a citation.

A compliance officer or any other employee of OSHA may not provide advance notice of the inspection under penalty of law.[7] The OSHA area director is, however, permitted to provide notice under the following circumstances:

---

[1] See e.g., Levin, *Crimes Against Employees: Substantive Criminal Sanctions Under the Occupational Safety and Health Act*, 14 Am. Crim. L. Rev., 98 (1977).
[2] See Chapter 5.
[3] See, *L.A. Law: Prosecuting Workplace Killers*, A.B.A. #J. 48, (Los Angeles prosecutor's *roll out* program could serve as model for OSHA.)
[4] *See infra*
[5] *See infra* sections 1.10 and 1.12.
[6] 29 C.F.R. Section 1903.8.
[7] 29 U.S.C. Section 17(f). The penalty for providing advance notice, upon conviction, is a fine of not more than $1000, imprisonment for not more than 6 months, or both.

1. In cases of apparent imminent danger, to enable the employer to correct the danger as quickly as possible
2. When the inspection can most effectively be conducted after regular business hours or where special preparations are necessary
3. To ensure the presence of employee and employer representatives or appropriate personnel needed to aid in inspections
4. When the area director determines that advance notice would enhance the probability of an effective and thorough inspection[1]

Compliance officers can also take environmental samples and obtain photographs related to the inspection. Additionally, compliance officers, can use other *reasonable investigative techniques*, including personal sampling equipment, dosimeters, air sampling badges, and other equipment.[2] Compliance officers must, however, take reasonable precautions when using photographic or sampling equipment to avoid creating hazardous conditions (i.e., a spark-producing camera flash in a flammable area) or disclosing a trade secret.[3]

An OSHA inspection has four basic components: (1) the opening conference, (2) the walk-through inspection, (3) the closing conference, and (4) the issuance of citations, if necessary. In the opening conference, the compliance officer may explain the purpose and type of inspection to be conducted, request records to be evaluated, question the employer, ask for appropriate representatives to accompany him or her during the walk-through inspection, and ask additional questions or request more information. The compliance officer may, but is not required to, provide the employer with copies of the applicable laws and regulations governing procedures and health and safety standards. The opening conference is usually brief and informal, its primary purpose is to establish the scope and purpose of the walk-through inspection.

After the opening conference and review of appropriate records, the compliance officer, usually accompanied by a representative of the employer and a representative of the employees, conducts a physical inspection of the facility or worksite.[4] The general purpose of this walk- through inspection is to determine whether the facility or worksite complies with OSHA standards. The compliance officer must identify potential safety and health hazards in the workplace, if any, and document them to support issuance of citations.[5]

---

[1] *Occupational Safety and Health Law*, 208-09 (1988).
[2] 29 C.F.R. Section 1903.7(b) [revised by 47 Fed. Reg. 5548 (1982)].
[3] See e.g., 29 C.F.R. Section 1903.9. Under section 15 of the OSHA Act, all information gathered or revealed during an inspection or proceeding that may reveal a trade secret as specified under 18 U.S.C. Section 1905 must be considered confidential, and breach of that confidentiality is punishable by a fine of not more than $1000, imprisonment of not more than one year, or both; and removal from office or employment with OSHA.
[4] It is highly recommended by the authors that a company representative accompany the OSHA inspection during the walk-through inspection.
[5] *OSHA Manual, supra* note 62, at III-D8.

The compliance officer uses various forms to document potential safety and health hazards observed during the inspection. The most commonly used form is the OSHA-1 Inspection Report where the compliance officer records information gathered during the opening conference and walk-through inspection.[1]

Two additional forms are usually attached to the OSHA-1 Inspection Report. The OSHA-1A form, known as the narrative, is used to record information gathered during the walk-through inspection: names and addresses of employees, management officials, and employee representatives accompanying the compliance officer on the inspection, and other information. A separate worksheet, known as OSHA-1B, is used by the compliance officer to document each condition that he or she believes could be an OSHA violation. One OSHA-1B worksheet is completed for each potential violation noted by the compliance officer.

When the walk-through inspection is completed, the compliance officer usually conducts an informal meeting with the employer or the employer's representative to "informally advise (the employer) of any apparent safety or health violations disclosed by the inspection."[2] The compliance officer informs the employer of the potential hazards observed and indicates the applicable section of the standards allegedly violated, advises that citations may be issued, and informs the employer or representative of the appeal process and rights.[3] Additionally, the compliance officer advises the employer that the OSHA Act prohibits discrimination against employees or others for exercising their rights.[4]

In an unusual situation, the compliance officer may issue a citation(s) on the spot. When this occurs the compliance officer informs the employer of the abatement period in addition to the other information provided at the closing conference. In most circumstances the compliance officer will leave the workplace and file a report with the area director who has authority, through the Secretary of Labor, to decide whether a citation should be issued, compute any penalties to be assessed; and set the abatement date for each alleged violation. The area director, under authority from the Secretary, must issue the citation with "reasonable promptness."[5] Citations must be issued

---

[1] The OSHA-1 Inspection Report Form includes the following: the establishment's name, inspection number, type of legal entity, type of business or plant, additional citations, names and addresses of all organized employee groups, the authorized representative of employees, the employee representative contacted, other persons contacted, coverage information (state of incorporation, type of goods or services in interstate commerce, etc.), date and time of entry, date and time that the walk-through inspection began, date and time closing conference began, date and time of exit, whether a follow-up inspection is recommended, the compliance officer's signature and date, the names of other compliance officers, evaluation of safety and health programs (checklist), closing conference checklist, and additional comments.
[2] 29 C.F.R. Section 1903.7(e).
[3] OSHA Manual, supra at n. 62, at III-D9.
[4] 29 U.S.C. Section 660(c)(1).
[5] Ibid. at section 658.

in writing and must describe in detail the violation alleged, including the relevant standard and regulation. There is a six-month statute of limitations and the citation must be issued or vacated within this time period. OSHA must serve notice of any citation and proposed penalty by certified mail, unless there is personal service, to an agent or officer of the employer.[1]

After the citation and notice of proposed penalty is issued, but before the notice of contest by the employer is filed, the employer may request an informal conference with the OSHA area director. The general purpose of the informal conference is to clarify the basis for the citation, modify abatement dates or proposed penalties, seek withdrawal of a cited item, or otherwise attempt to settle the case. This conference, as its name implies, is an informal meeting between the employer and OSHA. Employee representatives must have an opportunity to participate if they so request. Safety professionals should note that the request for an informal conference does not *stay* (delay) the 15-working-day period to file a notice of contest to challenge the citation.[2]

Under the OSHA Act, an employer, employee, or authorized employee representative (including a labor organization) is given 15 working days from when the citation is issued to file a *notice of contest*. If a notice of contest is not filed within 15 working days, the citation and proposed penalty become a final order of the Occupational Safety and Health Review Commission (OSHRC), and is not subject to review by any court or agency. If a timely notice of contest is filed in good faith, the abatement requirement is tolled (temporarily suspended or delayed) and a hearing is scheduled. The employer also has the right to file a petition for modification of the abatement period (PMA) if the employer is unable to comply with the abatement period provided in the citation. If OSHA contests the PMA, a hearing is scheduled to determine whether the abatement requirements should be modified.

When the notice of contest by the employer is filed, the Secretary must immediately forward the notice to the OSHRC, which then schedules a hearing before its administrative law judge (ALJ). The Secretary of Labor is labeled the *complainant*, and the employer the *respondent*. The ALJ may affirm, modify, or vacate the citation, any penalties, or the abatement date. Either party can appeal the ALJ's decision by filing a petition for discretionary review (PDR). Additionally, any member of the OSHRC may *direct review* any decision by an ALJ, in whole or in part, without a PDR. If a PDR is not filed and no member of the OSHRC directs a review, the decision of the ALJ becomes final in 30 days. Any party may appeal a final order of the OSHRC by filing a petition for review in the U.S. Court of Appeals for the circuit in which the violation is alleged to have occurred or in the U.S. Court of Appeals for the District of Columbia Circuit. This petition for review must be filed within 60 days from the date of the OSHRC's final order.

[1] *Fed. R. Civ. P.* 4(d)(3).
[2] 29 U.S.C. Section 659(a).

In a disaster situation where local or state prosecutors (i.e., district attorney or DA) respond to a fatality or disaster situation, safety professionals should exercise extreme caution and acquire legal representation prior to providing any information or documents. The prosecutor is usually investigating from a criminal prospective and would utilize the state criminal code as the basis for any charges. Safety professionals should preserve all constitutional rights including the right to remain silent and the right to counsel in these situations. As noted in the *Miranda warnings* provided by law enforcement after an arrest, "you have the right to remain silent, everything you say can and will be used against you in a court of law; you have the right to counsel…"

In summation, most safety professionals are *shell shocked* after a major disaster. They have attempted to perform many functions, under extreme duress, and have made life and death decisions. Often the safety professionals have been performing these functions for hours on end, are functioning on adrenaline, and are sleep deprived. A disaster has happened — emotions are high — this is not the time to compound the situation by making mistakes that would lead to additional long-term damage or life-altering penalties.

chapter eighteen

Legal issues

> "Laws are not invented; they grow out of circumstances."
>
> Azarias

> "Four out of five potential litigants will settle their disputes the first day they come together, if you will put the idea of arbitration into their heads."
>
> Moses H. Grossman

Immediately after a disaster situation, the *finger pointing* usually begins. Employees have been injured or killed, families have lost loved ones, property has been damaged, product has been lost, orders are not getting filled and *someone is going to be responsible.* Within a relatively short period following a disaster situation, safety professionals can expect to be inundated with claims, legal complaints, insurance adjusters, and a myriad of other legal issues of various types and various sources. Safety professionals should prepare for this deluge of legal matters and integrate the management of this area within the overall emergency and disaster preparedness program.

One of the major areas of legal concern following a disaster situation is the workers' compensation claims for employees who have been injured or killed. Safety professionals must have a firm grasp of the specific laws governing workers' compensation in their state and have established appropriate lines of communications with their insurance administrator or insurance carrier. Payment of death benefits in a timely manner for the families of deceased employees is essential. Acquisition of the appropriate medical care and payment of time-loss benefits for injured employees maintains the level of trust to minimize protracted litigation in the future. Additionally, the timely payment of workers' compensation benefits can often serve as a defense against additional litigation given the *sole remedy* provisions in many state workers' compensation laws.

In general, workers' compensation systems are fundamentally a no-fault mechanism through which employees, who incur work-related injuries and illnesses, are compensated with monetary and medical benefits. Either party's potential negligence is not an issue as long as this is the employer/employee relationship. In essence, workers' compensation is a compromise in that employees are guaranteed a percentage of their wages (generally 2/3) and full payment for their medical costs when injured on the job. Employers are guaranteed a reduced monetary cost for these injuries or illnesses and are provided a protection from additional or future legal action by the employee for the injury.

The typical workers' compensation system possesses the following features:

1. Every state in the U.S. has a workers' compensation system. There may be variations in the amounts of benefits, the rules, administration, etc. from state to state. In most states, workers' compensation is the exclusive remedy for on-the-job injuries and illnesses.
2. Coverage for workers' compensation is limited to employees who are injured on the job. The specific locations as to what constitutes the work premises and on the job may vary from state to state.
3. Negligence or fault by either party is largely inconsequential. No matter whether the employer is at fault or the employee is negligent, the injured employee generally receives workers' compensation coverage for any injury or illness incurred on the job.
4. Workers' compensation coverage is automatic, i.e., employees are not required to sign up for workers' compensation coverage. By law, employers are required to obtain and carry workers' compensation insurance or be self-insured.
5. Employee injuries or illnesses that *arise out of* and/or *are in the course of* employment are considered compensable. These definition phrases have expanded this beyond the four corners of the workplace to include work-related injuries and illnesses incurred on the highways, at various in and out-of-town locations, and other locals. These two concepts, *arising out of* the employment and *in the course of* the employment, are the basic burdens of proof for the injured employee. Most states require both. The safety and health professional is strongly advised to review the case law in his or her state to see the expansive scope of these two phrases. That is, the injury or illnesses must *arise out of*, i.e., there must be a causal connection between the work and the injury or illness must be *in the course of* the employment; this relates to the time, place, and circumstances of the accident in relation to the employment. The key issue is a **work connection** between the employment and the injury/illness.
6. Most workers' compensation systems include wage-loss benefits (sometimes known as time-loss benefits) which are usually between 1/2 to 3/4 of the employee's average weekly wage. These benefits

Chapter eighteen: Legal issues 117

are normally tax free and are commonly called temporary total disability (TTD) benefits.
7. Most workers' compensation systems require payment of all medical expenses, including such expenses as hospital expenses, rehabilitation expenses, and prothesis expenses.
8. In situations where an employee is killed, workers' compensation benefits for burial expenses and future wage-loss benefits are usually paid to the dependents.
9. When an employee incurs an injury or illness that is considered permanent in nature, most workers' compensation systems provide a dollar value for the percentage of loss to the injured employee. This is normally known as permanent partial disability (PPD) or permanent total disability (PTD).
10. In accepting workers' compensation benefits, the injured employee is normally required to waive any common law action to sue the employer for damages from the injury or illness.
11. If the employee is injured by a third party, the employer usually is required to provide workers' compensation coverage but can be reimbursed for these costs from any settlement that the injured employee receives through legal action or other methods.
12. Administration of the workers' compensation system in each state is normally assigned to a commission or board. The commission/board generally oversees an administrative agency located within state government which manages the workers' compensation program within the state.
13. The workers' compensation act in each state is a statutory enactment that can be amended by the state legislatures. Budgetary requirements are normally authorized and approved by the legislatures in each state.
14. The workers' compensation commission/board in each state normally develops administrative rules and regulations (i.e., rules of procedure, evidence, etc.) for the administration of workers' compensation claims in the state.
15. In most states, employers with one or more employees are normally required to possess workers' compensation coverage. Employers are generally allowed several avenues through which to acquire this coverage. Employers can select to acquire workers' compensation coverage from private insurance companies, from state-funded insurance programs, or become self-insured (i.e., after posting bond, the employer pays all costs directly from its coffers).
16. Most state workers' compensation provides a relatively long statute of limitations. For injury claims, most states grant between 1 and 10 years in which to file the claim for benefits. For work related illnesses, the statute of limitations may be as high as 20 to 30 years from the time the employee first noticed the illness or the illness was

diagnosed. An employee who incurred a work-related injury or illness is normally not required to be employed with the employer when the claim for benefits is filed.
17. Workers' compensation benefits are generally separate from the employment status of the injured employee. Injured employees may continue to maintain workers' compensation benefits even if the employment relationship is terminated, the employee is laid off, or other significant changes are made in the employment status.
18. Most state workers' compensation systems possess some type of administrative hearing procedures. Most workers' compensation acts have designed a system of administrative judges (normally known as administrative law judges or ALJ) to hear any disputes involving workers' compensation issues. Appeals of the decision of the administrative law judges are made normally to the workers' compensation commission/board. Some states permit appeals to the state court system after all administrative appeals have been exhausted.

Safety and loss prevention professionals should be very aware that the workers' compensation system in every state is administrative in nature. Thus, there is a substantial amount of required paperwork that must be completed in order for benefits to be paid in a timely manner. In most states, specific forms have been developed.

The most important form to initiate workers' compensation coverage in most states is the first report of injury/illness form. This form may be called a First Report form (an application for adjustment of claim) or may possess some other name or acronym like the SF-1 or Form 100. This form, often divided into three parts in order that information can be provided by the employer, employee, and attending physician, is often the catalyst that starts the workers' compensation system reaction. If this form is absent or misplaced, there is no reaction in the system and no benefits are provided to the injured employee.

Under most workers' compensation systems, there are many forms that need to be completed in an accurate and timely manner. Normally, specific forms must be completed if an employee is to be off work or is returning to work. These include forms for the transfer from one physician to another, forms for independent medical examinations, forms for the payment of medical benefits, and forms for the payment of permanent partial or permanent total disability benefits. Safety and loss prevention professionals responsible for workers' compensation are advised to acquire a working knowledge of the appropriate legal forms used in their state's workers' compensation program.

In most states, information regarding the rules, regulations, and forms can be acquired directly from the state workers' compensation commission/board. Other sources for this information include your insurance carrier, self-insured administrator, or state-fund administrator.

Chapter eighteen: Legal issues

Safety professionals should be aware that workers' compensation claims possess a *long tail*, i.e., stretch over a long period of time. Under the OSHA recordkeeping system, with which most safety professionals are familiar, every year injuries and illnesses are totaled on the OSHA Form 200 log and a new year begins. This is not the case with workers' compensation. Once an employee sustains a work-related injury or illness, the employer is responsible for the management and costs until such time as the injury or illness reaches maximum medical recovery or the time limitations are exhausted. When an injury reaches maximum medical recovery, the employer may be responsible for payment of permanent partial or permanent total disability benefits prior to closure of the claim. Additionally, in some states, the medical benefits can remain open indefinitely and cannot be settled or closed with the claim. In many circumstances, the workers' compensation claim for a work-related injury or illness may remain open for several years and thus require continued management and administration for the duration of the claim process.

Some states allow the employer to take the deposition of the employee claiming benefits, while others strictly prohibit it. Some states have a schedule of benefits and have permanent disability awards strictly on a percentage of disability from that schedule. Others require that a medical provider outline the percentage of functional impairment due to the injury/illness (see the American Medical Association (AMA) Guidelines), then using this information as well as the employee's age, education, and work history, the ALT determines the amount of occupational impairment upon which permanent disability benefits are awarded. Still other states have variations on those systems.

Safety prevention professionals who are responsible for the management of a workers' compensation program should become knowledgeable in the rules, regulations, and procedures under their individual state's workers' compensation system and be prepared to address a multitude of claims following a disaster situation. There is no substitute for knowing the rules and regulations under your state's workers' compensation system and being prepared as part of the overall emergency and disaster preparedness program.

In addition to workers' compensation claims, it is not unusual for wrongful death actions and third party liability actions to be filed against the company or the manufacturer of the equipment involved in the disaster situation. Safety professionals are often involved in these actions given their involvement and knowledge of the disaster situation. Safety professionals should always seek the assistance of their corporate legal department or other legal counsel if involved in these situations.

As discussed throughout this text, proper preparation is the key to appropriately managing the workers' compensation claims and litigation that follow a disaster situation. The safety manager cannot perform this function as well as all of the other important items involved in a disaster situation, thus this important area should be delegated to the appropriate parties possessing the necessary expertise. Below, please find several items to consider when addressing this important area in your emergency and disaster preparedness plan:

1. Who manages the workers' compensation claims for our company?
2. Who handles initiating and processing claims within the facility?
3. Who would be available to immediately travel to the disaster scene?
4. What process is utilized for death benefit claims?
5. Does your organization possess in-house legal counsel?
6. Do you possess open lines of communications with your insurance administrators? Workers' compensation administrators? Legal counsel?
7. Where would you house legal and claims administrators on site?
8. Do you have contracts with local psychologists and psychiatrists for critical stress debriefing?
9. How would you handle a claimant or attorney requesting to visit the disaster scene?
10. Who is responsible for handling all telephone inquiries?
11. Who is responsible for handling all legal service?
12. Who is responsible for all records, videotape, and other disaster scene documents?
13. What process is to be used to notify the families of injured or killed employees?
14. Who is responsible for addressing third-party actions against equipment manufacturers?
15. Who is responsible for coordinating the entire workers' compensation and legal arena?

Most safety professionals have just experienced a very stressful event during a disaster with many *life and death* decisions. The disaster event is now under control and the workers' compensation and legal issues begin to evolve. Appropriate and effective management of the claims and legal action from the beginning can result in effective and cost efficient results at the end of the *long tail* of the claims or litigation. The damage that can be caused by ineffective or non-existent management of these vital areas can be as damaging, or even more damaging, in the long run for the company or organization. Proper preparedness can ensure appropriate management of these important areas in the event of a disaster.

chapter nineteen

Disability issues

> "Courage and perseverance have a magical talisman before which difficulties disappear and obstacles vanish into air."
>
> John Quincy Adams

> "The great pleasure in life is doing what people say you cannot do."
>
> Walter Bagehot

An area often overlooked by safety professionals in the development of an emergency and disaster preparedness program is accommodations for individuals with disabilities. Prudent safety professionals are encouraged to specifically address the needs of individuals with disabilities as part of the overall planning process, especially in areas of alarms, egress routes, and related areas to ensure individuals with disabilities are able to participate within the overall emergency and disaster preparedness program.

Prudent safety professionals should know the requirements of the laws providing protections to individuals with disabilities. The most recent and most applicable to the emergency and disaster preparedness area is the Americans with Disabilities Act of 1990 (known as the ADA) which has opened new areas of regulatory compliance that will affect most emergency and disaster preparedness programs. The ADA prohibits discrimination against qualified individuals with physical or mental disabilities in all employment settings. Given the impact of the ADA on the job functions of employees especially in the areas of alarm systems, access and egress, facility modifications, and other areas, it is critical for safety professionals to possess a firm grasp of the scope and requirements of this new law.

From most estimates, the Americans with Disabilities Act has afforded protection to approximately 43 to 45 million individuals or approximately one in five Americans. In terms of the effect on the American workplace, the estimates of protected individuals when compared to the number of individuals currently employed in the American workplace (approximately 200 million), provide that employers can expect approximately one in four individuals currently employed or potential employees will be afforded protection under the ADA.

Structurally, the ADA is divided into five titles, and all titles possess the potential of substantially impacting EMS professionals in covered public or private sector organizations. Title I contains the employment provisions which protect all individuals with disabilities who are in the U.S., regardless of national origin and immigration status. Title II prohibits discriminating against qualified individuals with disabilities or excluding qualified individuals with disabilities from the services, programs, or activities provided by public entities. Title II contains the transportation provisions of the Act. Title III, entitled "Public Accommodations," requires that goods, services, privileges, advantages, and facilities of any public place be offered in the most integrated setting appropriate to the needs of the individual.

Title III also covers transportation offered by private entities. It addresses telecommunications. Title IV requires that telephone companies provide telecommunication relay services and television public service announcements produced or funded with federal money include closed caption. Title V includes the miscellaneous provisions. This title noted that the ADA does not limit or invalidate other federal and state laws providing equal or greater protection for the rights of individuals with disabilities and addresses related insurance, alternate dispute, and congressional coverage issues.

Title I of the ADA went into effect for all employers and industries engaged in interstate commerce with 25 or more employees on July 26, 1992. On July 26, 1994, the ADA became effective for all employers with 15 or more employees.[1] Title II, applies to public services such as emergency services and fire departments,[2] and Title III requiring public accommodations and services operated by private entities became effective on January 26, 1992,[3] except for specific subsections of Title II which went into effect immediately on July 26, 1990.[4] A telecommunication relay service required by Title IV was required to be available by July 26, 1993.[5]

Title I prohibits covered employers from discriminating against a *qualified individual with a disability* (QID) with regard to job applications, hiring, advancement, discharge, compensation, training, and other terms, conditions, and privileges of employment.[6]

> Section 101 (8) defines a *qualified individual with a disability* as any person who, with or without reasonable

---
[1] ADA Section 101(5), 108, 42 U.S.C. 12111.
[2] ADA Section 204(a), 42 U.S.C. 12134.
[3] Ibid.
[4] ADA Section 203 (a), 306 (a), 42 U.S.C. 12186.
[5] ADA Section 102 (a), 42 U.S.C. 12112.
[6] Ibid.

accommodation, can perform the essential functions of the employment position that such individual holds or desires...consideration shall be given to the employer's judgment as to what functions of a job are essential, and if an employer has prepared a written description before advertising or interviewing applicants for the job, this description shall be considered evidence of the essential function of the job.[1]

The Equal Employment Opportunity Commission (EEOC) provides additional clarification as to this definition in stating "an individual with a disability who satisfies the requisite skill, experience and educational requirements of the employment position such individual holds or desires, and who, with or without reasonable accommodation, can perform the essential functions of such position."[2]

Congress did not provide a specific list of disabilities covered under the ADA because of the difficulty of ensuring the comprehensiveness of such a list.[3] Under the ADA, an individual has a disability if he or she possesses:

1. A physical or mental impairment that substantially limits one or more of the major life activities of such individual,
2. A record of such an impairment, or
3. Is regarded as having such an impairment.[4]

For an individual to be considered disabled under the ADA, the physical or mental impairment must limit one or more major life activities. Under the U.S. Justice Department's regulations issued for section 504 of the Rehabilitation Act, *major life activities* is defined as, "functions such as caring for one's self, performing manual tasks, walking, seeing, hearing, speaking, breathing, learning and working."[5] Congress clearly intended to have the term *disability* construed broadly. However, this definition includes neither simple physical characteristics, nor limitations based on environmental, cultural, or economic disadvantages.[6]

Prudent safety professionals should also be aware that at this writing, the U.S. Supreme Court is reviewing several cases which may change the definition of disabled under the ADA. When the U.S. Supreme Court renders its decision, it is highly recommended that safety professionals acquire and review the decision closely.

---

[1] ADA Section 101 (8).
[2] EEOC Interpretive Rules, 56 Fed. Reg. 35 (July 26, 1991).
[3] 42 FR 22686 (May 4, 1977); S. Rep. 101-116; H. Rep. 101-485, Part 2, 51.
[4] Subtitle A, Section 3(2). The ADA departed from the Rehabilitation Act of 1973 and other legislation is using the term "disability" rather than "handicap".
[5] 28 C.F.R. Section 41.31. This provision is adopted by and reiterated in the Senate Report at page 22.
[6] *See Jasany v. U.S. Postal Service*, 755 F2d 1244 (6th Cir. 1985).

The second prong of this definition is "a record of such an impairment disability." The Senate Report and the House Judiciary Committee Report each stated:

> This provision is included in the definition in part to protect individuals who have recovered from a physical or mental impairment which previously limited them in a major life activity. Discrimination on the basis of such a past impairment would be prohibited under this legislation. Frequently occurring examples of the first group (i.e., those who have a history of an impairment) are people with histories of mental or emotional illness, heart disease or cancer; examples of the second group (i.e., those who have been misclassified as having an impairment) are people who have been misclassified as mentally retarded.[1]

The third prong of the statutory definition of a disability extends coverage to individuals who are "being regarded as having a disability." The ADA has adopted the *regarded as* test used for section 504 of the Rehabilitation Act:

> "Is regarded as having an impairment" means (A) has a physical or mental impairment that does not substantially limit major life activities but is treated ... as constituting such a limitation; (B) has a physical or mental impairment that substantially limits major life activities only as a result of the attitudes of others toward such impairment; (C) has none of the impairments defined (in the impairment paragraph of the Department of Justice regulations) but is treated ... as having such an impairment.[2]

Under the EEOC's regulations, this third prong covers three classes of individuals:

- Persons who have physical or mental impairments that do not limit a major life activity but who are nevertheless perceived by covered entities (employers, places of public accommodation) as having such limitations. (For example, an employee with controlled high blood pressure that is not, in fact, substantially limited, is reassigned to less strenuous work because of his employer's unsubstantiated fear that

---

[1] S. Rep. 101-116, 23; H. Rep. 101-485, Part 2, 52-3.
[2] 45 C.F.R. 84.3 (j)(2)(iv), quoted from H. Rep. 101-485, Part 3, 29; S. Rep. 101-116, 23:H. Rep. 101-485, Part 2, 53; *Also see School Board of Nassau County, Florida v. Arline*, 107 S. Ct. 1123 (1987)(leading case).

the individual will suffer a heart attack if he continues to perform strenuous work. Such a person would be regarded as disabled.)[1]
- Persons who have physical or mental impairments that substantially limit a major life activity only because of a perception that the impairment causes such a limitation. (For example, an employee has a condition that periodically causes an involuntary jerk of the head, but no limitations on his major life activities. If his employer discriminates against him because of the negative reaction of customers, the employer would be regarding him as disabled and acting on the basis of that perceived disability.)[2]
- Persons who do not have a physical or mental impairment but are treated as having a substantially limiting impairment. (For example, a company discharges an employee based on a rumor that the employee is HIV-positive. Even though the rumor is totally false and the employee has no impairment, the company would nevertheless be in violation of the ADA.)[3]

Thus, a qualified individual with a disability under the ADA is any individual who can perform the essential or vital functions of a particular job with or without the employer accommodating the particular disability. The employer is provided the opportunity to determine the essential functions of the particular job before offering the position through the development of a written job description. This written job description will be considered evidence as to which functions of the particular job are essential and which are peripheral. In deciding the essential functions of a particular position, the EEOC will consider the employer's judgment, whether the written job description was developed prior to advertising or beginning the interview process, the amount of time spent on performing the job, the past and current experience of the individual to be hired, relevant collective bargaining agreements and other factors.[4]

The EEOC defines the term *essential function* of a job as meaning "primary job duties that are intrinsic to the employment position the individual holds or desires" and precludes any marginal or peripheral functions which may be incidental to the primary job function."[5] The factors provided by the EEOC in evaluating the essential functions of a particular job include the reason the position exists, the number of employees available, and the degree of specialization required to perform the job.[6] This determination is especially important to managers who may be required to develop the written job descriptions or may be required to determine the essential functions of a given position.

[1] EEOC Interpretive Guidelines, 56 Fed. Reg. 35, 742 (July 26, 1991).
[2] S. Comm. on Lab. and Hum. Resources Rep. at 24; H. Comm. on Educ. and Lab. Rep. at 53; H. Comm. on Jud. Rep. at 30-31.
[3] 29 C.F.R. Section 1630.2(1).
[4] ADA, Title I, Section 101(8).
[5] EEOC Interpretive Rules, *supra*, note 9.
[6] Ibid.

Of particular concern to safety personnel is the treatment of the disabled individual, who, as a matter of fact or due to prejudice, is believed to be a direct threat to the safety and health of others in the workplace. To address this issue, the ADA provides that any individual who poses a direct threat to the health and safety of others that cannot be eliminated by reasonable accommodation may be disqualified from the particular job.[1] The term *direct threat* to others is defined by the EEOC as meaning "a significant risk of substantial harm to the health and safety of the individual or others that cannot be eliminated by reasonable accommodation."[2] The determining factors which managers should consider in making this determination include the duration of the risk, the nature and severity of the potential harm, and the likelihood the potential harm will occur.[3]

Additionally, managers should consider the EEOC's Interpretive Guidelines which state:

> [If] an individual poses a direct threat as a result of a disability, the employer must determine whether a reasonable accommodation would either eliminate the risk or reduce it to an acceptable level. If no accommodation exists that would either eliminate the risk or reduce the risk, the employer may refuse to hire an applicant or may discharge an employee who poses a direct threat.[4]

Safety professionals should note that Title I additionally provides that if an employer does not make reasonable accommodation for the known limitations of a qualified individual with disabilities, it is considered to be discrimination. Only if the employer can prove that providing the accommodation would place an undue hardship on the operation of the employer's business can discrimination be disproved.

Section 101 (9) defines a *reasonable accommodation* as "making existing facilities used by employees readily accessible to and usable by the qualified individual with a disability" and includes job restriction, part-time or modified work schedules, reassignment to a vacant position, acquisition or modification of equipment or devices, appropriate adjustments or modification of examinations, training materials, or policies, the provisions of qualified readers or interpreters and other similar accommodations for … the QID.[5]

The EEOC further defines *reasonable accommodation* as:

1. Any modification or adjustment to a job application process that enables a qualified individual with a disability to be considered for

---

[1] ADA, Section 103(b).
[2] EEOC Interpretive Guidelines.
[3] Ibid.
[4] 56 Fed. Reg. 35,745 (July 26, 1991); *Also see, Davis v. Meese*, 692 F. Supp. 505 (ED Pa. 1988)(Rehabilitation Act decision).
[5] ADA Section 101 (9).

the position such qualified individual with a disability desires, and which will not impose an undue hardship on the ... business; or
2. Any modification or adjustment to the work environment, or to the manner or circumstances which the position held or desired is customarily performed, that enables the qualified individual with a disability to perform the essential functions of that position and which will not impose an undue hardship on the ... business; or
3. Any modification or adjustment that enables the qualified individual with a disability to enjoy the same benefits and privileges of employment that other employees enjoy and does not impose an undue hardship on the ... business.[1]

In essence, the covered employer is required to make reasonable accommodations for any/all known physical or mental limitations of the qualified individual with a disability unless the employer can demonstrate that the accommodations would impose an undue hardship on the business or the particular disability directly affects the safety and health of the qualified individual with a disability or others. Included under this section is the prohibition against the use of qualification standards, employment tests, and other selection criteria that tend to screen out individuals with disabilities, unless the employer can demonstrate the procedure is directly related to the job function. In addition to the modifications to facilities, work schedules, equipment and training programs, that employers initiate an *informal interactive (communication) process* with the qualified individual to promote voluntary disclosure of specific limitations and restrictions by the qualified individual to enable the employer to make appropriate accommodations to compensate for the limitation.[2]

Job restructuring within the meaning of section 101(9)(B) means modifying a job such that a disabled individual can perform its essential functions. This does not mean, however, that the essential functions themselves must be modified.[3] Examples of job restricting may include:

- Eliminating nonessential elements of the job
- Redelegating assignments
- Exchanging assignments with another employee
- Redesigning procedures for task accomplishment
- Modifying the means of communication that are used on the job[4]

---

[1] EEOC Interpretive Guidelines.
[2] Ibid.
[3] *See Gruegging v. Burke*, 48 Fair Empl. Prac. Cas. (BNA) 140 (DDC 1987); *Bento v. ITO Corp.*, 599 F. Supp. 731 (DRI 1984).
[4] EEOC Interpretive Guidelines, 56 Fed. Reg. 35,744 (July 26, 1991); *Also see* Rehabilitation Act decisions including *Harrison v. March*, 46 Fair Empl. Prac. Cas. (BNA) 971 (WD Mo. 1988); *Wallace v. Veteran Admin.*, 683 F. Supp. 758 (D. Kan. 1988).

Section 101 (10)(A) defines undue hardship as "an action requiring significant difficulty or expense," when considered in light of the following factors:

1. Nature and cost of the accommodation,
2. The overall financial resources and workforce of the facility involved,
3. The overall financial resources, number of employees, and structure of the parent entity, and
4. The type of operation including the composition and function of the workforce, the administration and fiscal relationship between the entity and the parent.[1]

Section 102 (c)(1) of the ADA provides that the prohibition against discrimination through medical screening, employment inquiries, and similar scrutiny. EMS professionals should be aware that underlying this section was Congress' conclusion that information obtained from employment applications and interviews was often used to exclude individuals with disabilities — particularly those with so-called hidden disabilities such as epilepsy, diabetes, emotional illness, heart disease, and cancer — before their ability to perform the job was even evaluated.[2]

Under section 102(c)(2), safety professionals should be aware that conducting pre-employment physical examinations of applicants and asking prospective employees if they are qualified individuals with disabilities is prohibited. Employers are further prohibited from inquiring as to the nature or severity of the disability even if the disability is visible or obvious. Managers should be aware that individuals may ask whether any candidates for transfer or promotion who have a known disability whether he or she can perform the required tasks of the new position if the tasks are job related and consistent with business necessity. An employer is also permitted to inquire as to the applicant's ability to perform the essential job functions prior to employment. The employer should use the written job descriptions as evidence of the essential functions of the position.[3]

Medical personnel may require medical examinations only if the medical examination is specifically job related and is consistent with business necessity. Medical examinations are permitted only after the applicant with a disability has been offered the job position. The medical examination may be given before the applicant starts the particular job and the job offer may be conditioned on the results of the medical examination if all employees are subject to the medical examinations and information obtained from the medical examination is maintained in separate confidential medical files. Employers are permitted to conduct voluntary medical examinations for current

---

[1] ADA Section 101(10)(a).
[2] S. Comm. on Lab. and Hum. Resources Rep. at 38; H. Comm. on Jud. Rep. at 42.
[3] ADA, Title I, Section 102(C)(2).

employees as part of an on-going medical health program but again the medical files must be maintained separately and in a confidential manner.[1]

The ADA does not prohibit medical personnel from making inquiries or requiring medical examinations or *fit for duty* examinations when there is a need to determine whether an employee is still able to perform the essential functions of the job or where periodic physical examinations are required by medical standards or federal, state, or local law.[2] This may be an important issue for the triage stage of any emergency or disaster preparedness program.

Title II of the ADA is designed to prohibit discrimination against disabled individuals by public entities. This title covers the provision of services, programs, activities, and employment by public entities. A public entity under Title II includes:

- A state or local government,
- Any department, agency, special purpose district, or other instrumentality of a state or local government, and
- The National Railroad Passenger Corporation (Amtrak), and any commuter authority as this term is defined in section 103(8) of the Rail Passenger Service Act.[3]

Title II of the ADA prohibits discrimination in the area of ground transportation including buses, taxis, trains, and limousines. Air transportation is excluded from the ADA but is covered under the Air Carriers Access Act.[4] Covered organizations may be affected in the purchasing or leasing of new vehicles and in other areas such as the transfer of disabled individuals to the hospital or other facilities. Title II requires covered public entities to ensure that new vehicles are accessible to and usable by the qualified individual including individuals in wheelchairs. Thus, vehicles must be equipped with lifts, ramps, wheelchair space, and other modifications unless the covered public entity can justify that such equipment necessary is unavailable despite a good faith effort to purchase or acquire this equipment. Covered organizations may want to consider alternative methods to accommodate the qualified individual such as use of an ambulance service or other alternatives.

Title III of the ADA builds upon the foundation established by the Architectural Barriers Act and the Rehabilitation Act. This title basically extends the prohibitions that currently exist against discrimination in facility construction or financed by the federal government to apply to all privately operated public accommodations. Title III is focused on the accommodations in public facilities including such covered entities as retail stores, law offices,

---

[1] ADA Section 102(c)(2)(A).
[2] EEOC Interpretive Guidelines, 56 Fed. Reg. 35,751 (July 26, 1991). Federally mandated periodic examinations include such laws as the Rehabilitation Act, Occupational Safety and Health Act, Federal Coal Mine Health Act, and numerous transportation laws.
[3] ADA Section 201(1).
[4] S. Rep 101-116, 21; H. Rep 101-485, Part 2; Part 3, 26-27.

medical facilities, and other public areas. This section requires that goods, services, and facilities, of any public place to provide "in the most integrated setting appropriate to the needs of the qualified individual with a disability" except where the qualified individual with a disability may pose a direct threat to the safety and health of others which cannot be eliminated through modification of company procedures, practices, or policies. Prohibited discrimination under this section includes prejudice or bias against the qualified individual with a disability in the *full and equal enjoyment* of these services and facilities.[1]

The ADA makes it unlawful for public accommodations not to remove architectural and communication barriers from existing facilities and transportation barriers from vehicles where such removal is readily achievable.[2] This statutory language is new and is defined as "easily accomplished and able to be carried out without much difficulty or expense."[3] As an example, moving shelves to widen an aisle, lowering shelves to permit access, etc. The ADA also requires that when a commercial facility or other public accommodation is undergoing a modification that affects the access to a primary function area, specific alterations must be made to afford accessibility to the qualified individual with a disability.

Title III also requires auxiliary aids and services be provided for the qualified individual with a disability including, but not limited to, interpreters, readers, amplifiers, and other devices (not limited or specified under the ADA) to provide the qualified individual with a disability with an equal opportunity for employment, promotion, etc.[4] Congress did, however, provide that auxiliary aids and services need not be offered to customers, clients, and other members of the public if the auxiliary aid or service creates an undue hardship on the business. Managers may want to consider alternative methods of accommodating the qualified individual with a disability. This section also addresses modification of existing facilities to provide access to the qualified individual with a disability and requires all new facilities to be readily accessible and usable by the qualified individual with a disability.

Title IV requires all telephone companies to provide telecommunications relay service to aid the hearing and speech impaired individual with a disability. The Federal Communication Commission issued a regulation requiring implementation of this requirement by July 26, 1992 and established guidelines for compliance. This section also requires that all public service programming and announcements funded with federal monies be equipped with closed caption for the hearing impaired.[5]

---

[1] ADA Section 302.
[2] ADA Section 302(b)(2)(A)(iv).
[3] ADA Section 301 (9).
[4] ADA Section 3(1).
[5] Report of the House Committee on Energy and Commerce on the Americans With Disabilities Act of 1990, HR Rep. No. 485, 101st Cong., 2d Sess., (1990) (hereinafter cited as H. Comm. on Energy and Comm. Rep.); H. Comm. on Educ. and Lab. Rep., supra.; S. Comm. on Lab. and Hum. Resources Rep., *supra*.

Chapter nineteen: Disability issues

Title V assures that the ADA does not limit or invalidate other federal or state laws that provide equal or greater protection for the rights of individuals with disabilities. A unique feature of TITLE V is the miscellaneous provisions and the requirement of compliance to the ADA by all members of Congress and all federal agencies. Additionally, Congress required all state and local governments to comply with the ADA and permitted the same remedies against the state and local governments as any other organizations.[1]

Congress expressed its concern that sexual preferences could be perceived as a protected characteristic under the ADA or the courts could expand ADA's coverage beyond Congress' intent. Accordingly, Congress included section 511(b) which contains an expansive list of conditions which are not to be considered within the ADA's definition of disability. This list includes transvestites, homosexuals, and bisexuals. Additionally, the conditions of transsexualism, pedophilia, exhibitionism, voyeurism, gender identity disorders not resulting from physical impairment, and other sexual behavior disorders are not considered as a qualified disability under the ADA. Compulsive gambling, kleptomania, pyromania, and psychoactive substance use disorders from current illegal drug use are also not afforded protection under the ADA.[2]

Safety professionals should be aware that individuals extended protection under this section of the ADA include all individuals associated with or having a relationship to the qualified individual with a disability. This inclusion is unlimited in nature, including family members, individuals living together, and an unspecified number of others.[3] The ADA extends coverage to all individuals, thus the protection is provided to all individuals, legal or illegal, documented or undocumented, living within the boundaries of the U.S. regardless of their status.[4] Under section 102(b)(4), unlawful discrimination includes "excluding or otherwise denying equal jobs or benefits to a qualified individual because of the known disability of the individual with whom the qualified individual is known to have a relationship or association."[5] Thus, the protection afforded under this section is not limited to only family relationships, there appears to be no limits on the kinds of relationships or association afforded protection. Of particular note is the inclusion of unmarried partners of persons with AIDS or other qualified disabilities under this section.[6]

As with the OSHA Act, the ADA requires that employers post notices of the pertinent provisions of the ADA in an accessible format in a conspic-

---

[1] ADA Section 501.
[2] ADA, Section 511(a),(b); section 508. There is some indication that many of the conditions excluded from the disability classification under the ADA may be considered a covered handicap under the Rehabilitation Act. *See Rezza v. US Dept. of Justice*, 46 Fair Empl. Prac. Cas. (BNA) 1336 (ED Pa. 1988) (compulsive gambling); *Fields v. Lyng*, 48 Fair Empl. Prac. Cas. (BNA) 1037 (D. Md. 1988) (kleptomania).
[3] ADA Sections 102(b)(4) and 302(b)(1)(E).
[4] H. Rep. 101-485,Part 2, 51.
[5] ADA Section 102(b)(4).
[6] H. Rep. 101-485, Part 2, 61-62; Part 3, 38-39.

uous location within the employer's facilities. A prudent safety professional may wish to provide additional notification on the job applications and other pertinent documents.[1]

Under the ADA, it is unlawful for an employer to discriminate on the basis of disability against a qualified individual with a disability in all areas including:

1. Recruitment, advertising, and job application procedures,
2. Hiring, upgrading, promotion, award of tenure, demotion, transfer, layoff, termination, right to return from layoff, and rehiring,
3. Rate of pay or other forms of compensation and changes in compensation,
4. Job assignments, job classifications, organization structures, position descriptions, lines of progression, and seniority lists,
5. Leaves of absence, sick leave, or other leaves,
6. Fringe benefits available by virtue of employment, whether or not administered by the employer,
7. Selection and financial support for training including apprenticeships, professional meetings, conferences and other related activities, and selection for leave of absence to pursue training,
8. Activities sponsored by the employer including social and recreational programs, and
9. Any other term, condition, or privilege of employment.[2]

The EEOC has also noted that it is "unlawful ... to participate in a contractual or other arrangement or relationship that has the effect of subjecting the covered entity's own qualified applicant or employee with a disability to discrimination." This prohibition includes referral agencies, labor unions (including collective bargaining agreements), insurance companies, and others providing fringe benefits, and organizations providing training and apprenticeships.[3]

Safety professionals should note that the ADA possesses no recordkeeping requirements, has no affirmative action requirements, and does not preclude or restrict anti-smoking policies. Additionally, the ADA possesses no retroactivity provisions.

The ADA has the same enforcement and remedy scheme as Title VII of the Civil Rights Act of 1964 as amended by the Civil Rights Act of 1991. Compensatory and punitive damages (with upper limits) have been added as remedies in cases of intentional discrimination, and there is a correlative right to a jury trial. Unlike Title VII, there is an exception where there exists good faith effort at reasonable accommodation.[4]

---

[1] ADA Section 105.
[2] EEOC Interpretive Guidelines.
[3] Ibid.
[4] Civil Rights Act of 1991, section 102.

The enforcement procedures adopted by the ADA mirror those of Title VII of the Civil Rights Act. A claimant under the ADA must file a claim with the EEOC within 180 days from the alleged discriminatory event or within 300 days in states with approved enforcement agencies such as the Human Rights Commission. These are commonly called dual agency states or section 706 agencies. The EEOC has 180 days to investigate the allegation and to sue the employer or issue a right-to-sue notice to the employee. The employee will have 90 days to file a civil action from the date of this notice.[1]

The original remedies provided under the ADA included reinstatement, with or without back pay, and reasonable attorney fees and costs. The ADA also provided for protection against retaliation against the employee for filing the complaint or others who may assist the employee in the investigation of the complaint. The ADA remedies are designed, as with the Civil Rights Act, to make the employee *whole* and to prevent future discrimination by the employer. All rights, remedies, and procedures of section 505 of the Rehabilitation Act of 1973 are also incorporated into the ADA. Enforcement of the ADA is also permitted by the Attorney General or by private lawsuit. Remedies under these titles included ordered modification of a facility, and civil penalties up to $50,000 for the first violation and $100,000 for any subsequent violations. Section 505 permits reasonable attorney fees and litigation costs for the prevailing party in an ADA action, but under section 513 Congress encourages the use of arbitration to resolve disputes arising under the ADA.[2]

With the passage of the Civil Rights Act of 1991, the remedies provided under the ADA were modified. Damages for employment discrimination, whether intentional or by practice which has a discriminatory effect, may include hiring, reinstatement, promotion, back pay, front pay, reasonable accommodation, or other action that will make an individual whole. Payment of attorneys' fees, expert witness fees and court courts were still permitted and jury trials were allowed.

Compensatory and punitive damages were also made available where intentional discrimination is found. Damages may be available to compensate for actual monetary losses, future monetary losses, mental anguish, and inconvenience. Punitive damages are also available if an employer acted with malice or reckless indifference. The total amount of punitive damages and compensatory damages for future monetary loss and emotional injury for each individual is limited, based upon the size of the employer. Punitive damages are **not** available against state or local governments.

Number of employees	Damages will not exceed
15–100	$ 50,000
101–200	100,000
201–500	200,000
500 or more	300,000

[1] S. Rep. 101-116, 21; H. Rep. 101-485 Part 2, 51; Part 3, 28.
[2] ADA Sections 505 and 513.

In situations involving reasonable accommodation, compensatory or punitive damages may not be awarded if the employer can demonstrate that "good faith" efforts were made to accommodate the individual with a disability.

Safety professionals should be aware that the Internal Revenue Code may provide tax credits and/or tax deductions for expenditures incurred in achieving compliance with the ADA. Programs like the Small Business Tax Credit and Targeted Job Tax Credit may be available upon request by the qualified employers. Additionally, expenses incurred in achieving compliance may be considered a deductible expense or capital expenditure permitting depreciation over a number of years under the Internal Revenue Code.

Title I — Employment provisions

The two-threshold questions often asked by safety professionals are whether the organization must comply with the ADA and who is a protected individual under the ADA. They are vitally important questions that must be addressed by managers in order to ascertain whether compliance is mandated and, if so, whether current employees, job applicants, and others who may directly affect the operation are within the protective scope of the ADA.

Question 1: Who must comply with Title I of the ADA?

Title I covers all private sector employers that affect commerce; state, local, and territorial governments, employment agencies, labor unions, and joint labor-management committees fall within the scope of a *covered entity* under the ADA.[1] Additionally, Congress and its agencies are covered, they are permitted to enforce the ADA through internal administrative procedures.[2] The federal government, government-owned corporations, Indian tribes, and tax-exempt private membership clubs (other than labor organizations who are exempt under section 501(c) of the Internal Revenue Code) are excluded from coverage under the ADA.[3]

Covered employers cannot discriminate against qualified applicants and employees on the basis of disability. Congress did provide a time period to enable employers to achieve compliance with the Title I. Coverage for Title I is phased in two steps based on the number of employees in order to allow additional time to smaller employers.

Number of employees	Effective date
25 or more employees	July 26, 1992
15 or more employees	July 26, 1994

---

[1] ADA Section 101(2) and 42 U.S.C. 12111.
[2] ADA Sections 509(a)(1), (b),(c)(2) and 42 U.S.C. 12209.
[3] ADA Section 101(5)(B) and 42 U.S.C. 12111.

State and local governments, regardless of size, are covered by employment nondiscrimination requirements under Title II of the ADA and had to be in compliance as of January 26, 1992. Certain individuals appointed by elected officials of state and local governments are covered by the same special enforcement procedures as established for Congress.

Similar to the coverage requirements under Title VII of the Civil Rights Act of 1964, an *employer* is defined to include persons who are agents of the employer such as safety and health managers, supervisors, personnel managers, and others. Thus, the corporation or legal entity which is the employer is responsible for the acts and omissions of their managerial employees and other agents who may violate the provisions of the ADA.

The second threshold question after an employer has ascertained that his or her organization or company is a covered entity required to comply with the ADA is which individuals are qualified for protection under the Title I and how are they identified as protected individuals? This question can be answered by asking the following questions:

- Who is protected by Title I?
- What constitutes a disability?
- Is the individual specifically excluded from protection under the ADA?

Question 2: Who is protected by Title I?

The ADA prohibits employment discrimination against qualified individuals with disabilities in such areas as job applications, hiring, testing, job assignments, evaluations, disciplinary actions, medical examinations, layoff/recall, discharge, compensation, leave, promotion, advancement, compensation, benefits, training, social activities, and other terms, conditions, and privileges of employment. A qualified individual with a disability is defined as:

> an individual with a disability who meets the skill, experience, education, and other job-related requirements of a position held or desired, and who, with or without reasonable accommodation, can perform the essential functions of a job.[1] Additionally, unlawful discrimination under the ADA includes: excluding or otherwise denying equal jobs or benefits to a qualified individual because of the known disability of an individual with whom the qualified individual is known to have a relationship or association.

This clause is designed to protect individuals who possess no disability themselves but who may be discriminated against because of their association or relationship to a disabled person. The protection afforded under this

---

[1] *Technical Assistance Manual for the Americans With Disabilities Act*, EEOC at 1-3.

clause is not limited to family members or relatives but extends in an apparently unlimited fashion to all associations and relationships. However, in an employment setting, if an employee is hired and then violates the employer's attendance policy, the ADA will not protect the individual from appropriate disciplinary action. The employer does not owe accommodation duty to an individual who is not disabled.

Question 3: What constitutes a disability?

Section 3(2) of the ADA provides a three-prong definition to ascertain who is and is not afforded protection. A person with a disability is an individual who:

> Test 1 — has a physical or mental impairment that substantially limits one or more of his or her major life activities,
> Test 2 — has a record of such an impairment; or
> Test 3 — is regarded as having such an impairment.

This definition is comparable to the definition of *handicap* under the Rehabilitation Act of 1973. Congress adopted this terminology in an attempt to use the most current acceptable terminology but intended that the relevant case law developed under the Rehabilitation Act be applicable to the definition of *disability* under the ADA.[1] It should be noted, however, that the definition and regulations applying to disability under the ADA are more favorable to the disabled individual than the handicap regulations under the Rehabilitation Act.

The first prong of this definition includes three major subparts that further define who is a protected individual under the ADA. These subparts, namely (1) a physical or mental impairment, (2) that substantially limits, (3) one or more of his or her major life activities, provide additional clarification as to the definition of a disability under the ADA.

A physical or mental impairment

The ADA does not specifically list all covered entities. Congress noted that:

> it is not possible to include in the legislation a list of all the specific conditions, diseases, or infections that would constitute physical or mental impairments because of the difficulty in ensuring the comprehensiveness of such a list, particularly in light of the fact that new disorders may develop in the future.[2]

---

[1] EEOC Interpretive Guidelines, 56 Fed. Reg. 35,740 (July 26, 1991); Report of the Senate Comm. on Labor and Human Resources on the Americans With Disabilities Act of 1989, S. Rep. No. 116, 101st Cong., 1st Sess. (1989).
[2] S. Comm. on Lab. and Hum. Resources Rep. at 22.

Chapter nineteen: Disability issues

A *physical impairment* is defined by the ADA as:

> any physiological disorder, or condition, cosmetic disfigurement, or anatomical loss affecting one or more of the following body systems: neurological, musculoskeletal, special sense organs, respiratory (including speech organs), cardiovascular, reproductive, digestive, genitalurinary, hemic and lymphatic, skin, and endocrine.[1]

A *mental impairment* is defined by the ADA as:

> any mental or psychological disorder, such as mental retardation, organic brain syndrome, emotional or mental illness, and specific learning disabilities.[2]

A person's impairment is determined without regard to any medication or assisting devices that the individual may use. For example, an individual with epilepsy who uses medication to control the seizures or a person with an artificial leg would be considered to have an impairment even if the medicine or prosthesis reduced the impact of the impairment.

The legislative history is clear that an individual with AIDS or HIV is protected by the ADA.[3] A contagious disease such as tuberculosis would also constitute an impairmment. However, an employer does not have to hire or retrain a person with a contagious disease if it poses a direct threat to the health and safety of others. This is discussed in detail later in this section.

The physiological or mental impairment must be permanent in nature. Pregnancy is considered temporary and thus is not afforded protection under the ADA, but is protected under other federal laws. Simple physical characteristics, such as hair color, left handedness, height, or weight within the normal range, are not considered impairments. Predisposition to a certain disease is not an impairment within this definition. Environmental, cultural, or economic disadvantages, such as lack of education or prison records, are not impairments. Similarly, personality traits such as poor judgment, quick temper, or irresponsible behavior are not impairments. Conditions such as stress and depression may or may not be considered an impairment depending on whether the condition results from a documented physiological or mental disorder.[4]

Case law under the Rehabilitation Act, applying similar language as in the ADA, has identified the following as some of the protected conditions: blindness, diabetes, cerebral palsy, learning disabilities, epilepsy, deafness, cancer, multiple sclerosis, allergies, heart conditions, high blood pressure, loss of leg, cystic fibrosis, hepatitis B, osteoarthritis, and many other conditions.

---

[1] *Technical Assistance Manual, supra.*
[2] Ibid.
[3] H.Comm. on Educ. and Lab. Rep. at 52; S.Comm. on Lab. and Hum. Resources Comm. at 22, 136 Cong. Rec. S9697 (July 13, 1990). *See also Technical Assistance Manual, supra.*
[4] *Technical Assistance Manual, supra.*

Substantial limits

While Congress clearly desired to have the term *disability* construed broadly, merely possessing an impairment is not sufficient for protection under the ADA. An impairment is only a disability under the ADA if it substantially limits one or more major life functions. An individual must be unable to perform or be significantly limited in performance in a basic activity that can be performed by an average person in America.

To assist in this evaluation, a three-factor test was provided to determine whether an individual's impairment substantially limits a major life activity:

- The nature and severity of the impairment
- How long the impairment will last or is expected to last
- The permanent and long-term impact, or expected impact of the impairment

The determination of whether an individual is substantially limited in a major life activity must be made on a case-by-case basis. The three-factor test should be considered because it is not the name of the impairment or condition that determines whether an individual is protected, but rather the effect of the impairment or condition on the life of the person. Some impairments such as blindness, AIDS, and deafness are by their nature substantially limiting, but other impairments may be disabling for some individuals and not for others, depending on the nature of the impairment and the particular activity.[1]

Individuals with two or more impairments, neither of which by itself substantially limits a major life activity, may be combined together to impair one or more major life activities. Temporary conditions such as a broken leg, common cold, sprains/strains are generally not protected because of the extent, duration, and impact of the impairment. However, such temporary conditions may evolve into a permanent condition, which substantially limits a major life function if complications arise.

In general, it is not necessary to determine if an individual is substantially limited in a work activity if the individual is limited in one or more major life activities. An individual is not considered to be substantially limited in working if he or she is substantially limited in performing only a particular job or unable to perform a specialized job in a particular area. An individual may be considered substantially limited in working if the individual is restricted in his or her ability to perform either a class of jobs or a broad range of jobs in various classes when compared to an average person of similar training, skills, and abilities. Factors to be considered include:

- The type of job from which the individual has been disqualified because of the impairment
- The geographical area in which the person may reasonably expect to find a job

---

[1] Ibid.

- The number and types of jobs using similar training, knowledge, skill, or abilities from which the individual is disqualified within the geographical area
- The number and types of other jobs in the area that do not involve similar training, knowledge, skill, or abilities from which the individual also is disqualified because of the impairment[1]

In evaluating the number of jobs from which an individual might be excluded, the EEOC regulations note that it is only necessary to show the approximate number of jobs from which the individual would be excluded.

Major life activities

An impairment must substantially limit one or more major life activities to be considered a disability under the ADA. A major life activity is an activity that an average person can perform with little or no difficulty. Examples include walking, seeing, speaking, hearing, breathing, learning, performing manual tasks, caring for oneself, standing, working, lifting, reading, and sitting.[2] This list of examples is not all-inclusive. All situations should be evaluated on a case-by-case basis.

The second test of this definition of disability requires that an individual possess a record of having an impairment as specified in Test 1. Under this test, the ADA protects individuals who possess a history of, or who have been misclassified as possessing a mental or physical impairment that substantially limits one or more major life functions. A record of impairment would include such documented items as educational, medical, or employment records. Safety and loss prevention professionals should note that merely possessing a record of being a *disabled veteran* or record of disability under another federal or state program does not automatically qualify the individual for protection under the ADA. The individual must meet the definition of *disability* under Test 1 and possess a record of such disability under Test 2.

The third test of the definition of disability includes an individual who is regarded or treated as having a covered disability even though the individual does not possess a disability as defined under Tests 1 and 2. This part of the definition protects individuals who do not possess a disability that substantially limits a major life activity from the discriminatory actions of others because of their perceived disability. This protection is necessary because society's myths and fears about disability and disease are as handicapping as are the physical limitations that flow from actual impairment.[3]

Three circumstances in which protection would be provided to the individual include:

---

[1] Ibid.
[2] Ibid.
[3] Ibid.

1. When the individual possesses an impairment which is not substantially limiting but the individual is treated by the employer as having such an impairment
2. When an individual has an impairment that is substantially limiting because of the attitude of others toward the condition
3. When the individual possesses no impairment but is regarded by the employer as having a substantially limiting impairment[1]

To acquire the protection afforded under the ADA, an individual must not only be an individual with a disability, but also must qualify under the above noted tests. A *qualified individual with a disability* is defined as a person with a disability who:

> satisfies the requisite skills, experience, education, and other job-related requirements of the employment position such individual holds or desires, and who, with or without reasonable accommodation, can perform the essential functions of such position.[2]

Managers should be aware that the employer is not required to hire or retain an individual who is not qualified to perform a particular job.

Question 4: Is the individual specifically excluded from protection under the ADA?

The ADA specifically provides a provision which excluded certain individuals from protection. As set forth under sections 510 and 511(a),(b), the following individuals are not protected:

- Individuals who are currently engaged in the use of illegal drugs are not protected when an employer takes action due directly to their continued use of illegal drugs. This includes illegal use of prescription drugs as well as illegal drugs. However, individuals who have undergone a qualified rehabilitation program and are not currently using drugs illegally are afforded protection under the ADA.
- Homosexuality and bisexuality are not impairments and therefore are not considered disabilities under the ADA.
- The ADA does not consider transvestism, transsexualism, pedophilia, exhibitionism, voyeurism, gender identity disorders not resulting from physical impairment, and other sexual behavior traits as disabilities and thus are not afforded protection.
- Other areas not afforded protection include compulsive gambling, kleptomania, pyromania, and psychoactive substance use disorders resulting from illegal use of drugs.

[1] EEOC Interpretive Guidelines, 56 Fed. Reg. 35, 744 (July 26, 1991).
[2] Title I, section 101.

Chapter nineteen: Disability issues

A major component of Title I is the *reasonable accommodation* mandate which requires employers to provide a disabled employee or applicant with the necessary reasonable accommodations that would allow the disabled individual to perform the essential functions of a particular job. Safety professionals should note that *reasonable accommodation* is a key nondiscrimination requirement in order to permit individuals with disabilities to overcome unnecessary barriers which could prevent or restrict employment opportunities.

The EEOC regulations define reasonable accommodation as meaning:

1. Any modification or adjustment to a job application process that enables a qualified individual with a disability to be considered for the position such qualified individual with a disability desires, and which will not impose an undue hardship on the … business, or
2. Any modification or adjustment to the work environment, or to the manner or circumstances which the position held or desired is customarily performed, that enables the qualified individual with a disability to perform the essential functions of that position and which will not impose an undue hardship on the … business, or
3. Any modification or adjustment that enables the qualified individual with a disability to enjoy the same benefits and privileges of employment that other employees enjoy and does not impose an undue hardship on the … business.[1]

Section 101(9) of the ADA states that reasonable accommodation includes two components. First, there is the accessibility component which sets forth an affirmative duty for the employer to make physical changes in the workplace in order that the facility is readily accessible and usable by individuals with disabilities. This component includes both those areas that must be accessible for the employee to perform the essential job functions, as well as non-work areas used by the employer's employees for other purposes.[2] The second component is modification of other related areas. The EEOC regulations set forth a number of examples of modification that an employer must consider:

- Job restructuring
- Part-time or modified work schedules
- reassignment to vacant position, appropriate adjustment or modification of examinations, training materials
- Acquisition or modification of equipment or devices
- Providing of qualified readers or interpreters[3]

Managers should note that the employer possesses no duty to make an accommodation for an individual who is not otherwise qualified for a position.

---
[1] EEOC Regs. at 29 C.F.R. Section 1630.2(o)(1).
[2] EEOC Interpretive Guidelines, 56 Fed. Reg. 35, 744 (July 26, 1991).
[3] Title I Section 101(9)(B); EEOC Regs. at 29 C.F.R. Section 1630.2(n)(2).

In most circumstances, it is the obligation of the individual with a disability to request a reasonable accommodation from the employer. The individual with a disability possesses the right to refuse an accommodation, but if the individual with a disability cannot perform the essential functions of the job without the accommodation, the individual with a disability may not be qualified for the job.

An employer is not required to make a reasonable accommodation that would impose an undue hardship on the business.[1] An undue hardship is defined as an action that would require significant difficulty or expense in relation to the size of the employer, the employer's resources, and the nature of the operations. Although the undue hardship limitations will be analyzed on a case-by-case basis, several factors have been set forth to determine whether an accommodation would impose an undue hardship. First, the undue hardship limitation would be unduly costly, extensive or substantial in nature, disruptive to the operation, or the accommodation would fundamentally alter the nature or operation of the business.[2] Additionally, the ADA provides four factors to be considered in determining whether an accommodation would impose an undue hardship on a particular operation:

1. The nature and the cost of the accommodation needed
2. The overall financial resources of the facility or facilities making the accommodation, the number of employees in the facility, and the effect on expenses and resources of the facility
3. The overall financial resources, size, number of employees, and type and location of facilities of the entity covered by the ADA
4. The type of operation of the covered entity, including the structure and functions of the workforce, the geographic separateness, and the administrative or fiscal relationship of the facility involved in making the accommodation to the larger entity[3]

Other factors such as the availability of tax credits and tax deductions, the type of enterprise, etc., can also be considered when evaluating an accommodation situation for the undue hardship limitation. Safety professionals should note that the requirements to prove undue hardship are substantial in nature and cannot easily be utilized to circumvent the purposes of the ADA.

The ADA prohibits the use of pre-employment medical examinations, medical inquiries, and requests for information regarding workers' compensation claims prior to an offer of employment.[4] An employer, however, may condition a job offer (i.e., conditional or contingent job offer) on the satisfactory results of a post-offer medical examination if the medical examination is required of all applicants or employees in the same job classification.

[1] Title I Section 101(10(a); *EEOC Technical Assistance Manual, supra.*
[2] *EEOC Technical Assistance Manual, supra.*
[3] Title I Section 101(10(B); *EEOC Technical Assistance Manual, supra.*
[4] Title I Section 102(c)(1).

Chapter nineteen: Disability issues

Questions regarding other injuries and workers' compensation claims may also be asked following the offer of employment. A post-offer medical examination cannot be used to disqualify an individual with a disability who is currently able to perform the essential functions of a particular job because of speculation that the disability may cause future injury or workers' compensation claims.

Safety professionals should note that if an individual is not employed because the medical examination revealed a disability, the reason for not hiring the qualified individual with a disability must be business related and necessary for the particular business. The burden of proving that a reasonable accommodation would not have enabled the individual with a disability to perform the essential functions of the particular job or the accommodation was unduly burdensome falls squarely on the employer.

As often revealed in the post-offer medical examination, the physician should be informed that the employer possesses the burden of proving that a qualified individual with a disability should be excluded because of the risk to the health and safety of other employees or individuals. To address this issue, Congress specifically noted that the employer may possess a job requirement that specified an individual not impose a direct threat to the health and safety of other individuals in the workplace.[1] A *direct threat* has been defined as meaning a significant risk to the health and safety of others that cannot be eliminated or reduced by reasonable accommodation.[2]

Managers should be aware that the direct threat evaluation is vitally important in evaluating disabilities involving contagious diseases. The leading case in this area is *School Board of Nassau County v. Arline*.[3] This case sets forth the test to be used in evaluating a direct threat to others.

- The nature of the risk
- The duration of the risk
- The severity of the risk
- The probability the disease will be transmitted and will cause varying degrees of harm[4]

The ADA imposes a very strict limitation on the use of information acquired through post-offer medical examination or inquiry. All medical-related information must be collected and maintained on separate forms and kept in separate files. These files must be maintained in a confidential manner with appropriate security and only designated individuals provided access. Medical-related information may be shared with appropriate first aid and safety personnel when applicable in an emergency situation. Supervisors and other managerial personnel may be informed about necessary job restrictions or job accommodations. Appropriate insurance organizations may

---
[1] Title I Section 103(b).
[2] Title I Section 101(c) and 29 CFR Section 1530.2(r).
[3] 480 US 273 (1987).
[4] Ibid.

acquire access to medical records when required for health or life insurance. State and federal officials may acquire access to medical records for compliance and other purposes.

In the area of insurance, the ADA specifies that nothing within the Act is to be construed to prohibit or restrict an insurer, hospital, or medical service company, health maintenance organization, or any agent, or entity that administers benefit plans, or similar organization from underwriting risks, or administering such risks that are based on or are not inconsistent with state laws.[1] An employer may not classify or segregate an individual with a disability in a manner that adversely affects not only the individual's employment but any provisions or administration of health insurance, life insurance, pension plans, or other benefits. In essence, this means that if an employer provides insurance or benefits to all employees, the employer must also provide this coverage to the individual with a disability. An employer cannot deny insurance to or subject the individual with a disability to different terms or conditions of insurance based upon the disability alone if the disability does not pose an increased insurance risk. An employer cannot terminate or refuse to hire an individual with a disability because the individual's disability or a family member or dependent's disability is not covered under his current policy or because the individual poses a future risk of increased health costs. The ADA does not, however, prohibit the use of preexisting condition clauses in insurance policies.

An employer is prohibited from shifting away the responsibilities and potential liabilities under the ADA through contractual or other arrangements. An employer may not do anything through a contractual relationship that it cannot do directly.[2] This provision applies to all contractual relationships that include insurance companies, employment and referral agencies, training organizations, agencies used for background checks, and labor unions.

Labor unions are covered by the ADA and have the same responsibilities as any other covered employer. Employers are prohibited from taking any action through a collective bargaining agreement (i.e., union contract) that it may not take directly by itself. A collective bargaining agreement may be used as evidence in a decision regarding undue hardship and in identifying the essential elements in a job description.

Although not required under the ADA, a written job description describing the essential elements of a particular job is the first line of defense for most ADA related claims. A written job description that is prepared before advertising or interviewing applicants for a job will be considered as evidence of the essential elements of the job along with other relevant factors.

In order to identify the essential elements of a particular job and thus whether an individual with a disability is qualified to perform the job, the EEOC regulations set forth three key factors, among others, which must be considered:

---

[1] H. Comm. on Educ. and Lab. Rep. at 59, 137.
[2] Title I section 102(b)(2); *EEOC Technical Assistance Manual, supra*.

- The reason the position exists to perform the function
- The limited number of employees available to perform the function, or among whom the function can be distributed
- The task function is highly specialized, and the person in the position is hired for special expertise or ability to perform the job

Title II — Public services

Title II is designed to prohibit discrimination against disabled individuals by public entities. Title II covers all services, programs, activities, and employment by government or governmental units. Title II adopted all of the rights, remedies, and procedures provided under section 505 of the Rehabilitation Act of 1973 and the undue financial burden exception is applicable.[1] The effective date for Title II was January 26, 1992.

The public entities which Title II applies includes state or local government, any department, agency, special purpose district, or other instrumentality of a state or local government, and the National Railroad Passenger Corporation (Amtrak) and any commuter authority as defined in the Rail Passenger Service Act.[2]

Title II possesses two basic purposes, namely to extend the prohibition against discrimination under the Rehabilitation Act of 1973 to state and local governments, and to clarify section 504 of the Rehabilitation Act for public transportation entities that receive federal assistance.[3] Given these purposes, the main emphasis of Title II is directed at the public sector organizations and possesses minimal impact on the private sector organizations.

The vast majority of Title II's provisions cover transportation provided by public entities to the general public such as buses and trains. The major requirement under Title II mandates that public entities who purchase or lease new buses, rail cars, taxis, or other vehicles must assure that these vehicles are accessible to and usable by qualified individuals with disabilities. This accessibility requirement includes disabled individuals who may be wheelchair bound and requires that all vehicles be equipped with lifts, ramps, wheelchair spaces, or other special accommodations unless the public entity can prove such equipment is unavailable despite a good faith effort to locate the equipment.

Many public entities purchase used vehicles or lease vehicles due to the substantial cost of such vehicles. The public entity must make a good-faith effort to obtain vehicles which are readily accessible and usable by individuals with disabilities. As provided under the ADA, it is considered discrimination to remanufacture vehicles to extend their useful life for five years or more without making the vehicle accessible and usable by individuals with disabilities. Historical vehicles, such as the trolley cars, may be excluded if

[1] 29 U.S.C. Section 794.
[2] Title II Section 103(8).
[3] H. Comm. on Educ. and Lab. Rep. at 84.; S. Comm. on Lab. and Hum Resources Rep. at 44; H. Comm. on Energy and Comm. Rep. at 26.

the modification to make the vehicle readily accessible and usable by individuals with disabilities alters the historical character of the vehicle.

Of particular importance to police, fire, and other emergency organizations is Title II's impact on 911 systems. Congress observed that many 911 telephone numbering systems were not directly accessible to hearing impaired and speech impaired individuals.[1] Congress cited, as an example, a deaf woman who died of a heart attack because the police organization did not respond when her husband tried to use his telephone communication device for the deaf (TTD) to call 911.[2] In response to such examples, Congress stated, "As part of its prohibition against discrimination in local and state programs and services, Title II will require local governments to ensure that these telephone emergency number systems (911) are equipped with technology that will give hearing impaired and speech impaired individuals a direct line to these emergency services.[3] Thus, public safety organizations had to ensure compliance with this requirement as of January 26, 1992.

Of importance for state governments is the fact that section 502 eliminates immunity of a state in state or federal court under the Eleventh Amendment for violations of the ADA. A state can be found liable in the same manner and is subject to the same remedies, including attorney fees, as private sector covered organizations.

Additionally, the claims procedures for instituting a complaint against a state or local government is significantly different than against a private covered entity. The ADA provides that a claim can be filed with any of seven federal government agencies including the EEOC and the Justice Department, or EEOC may assist in such litigation. A procedure for instituting complaints against a public organization without going to court is provided in the Justice Department's regulations. The statute of limitations on filing such a claim with the designated federal agency is 180 days from the date of the act of discrimination, unless the agency extends the time limitation for good cause. If the responsible agency finds a violation, the violation will be corrected through voluntary compliance, negotiations, or intervention by the Attorney General.

This procedure is totally voluntary. An individual may file suit in court without filing an administrative complaint, or an individual may file suit at any time while an administrative complaint is pending. No exhaustion of remedies is required.[4]

Under the Department of Justice's regulations, public entities with fifty or more employees are required to designate at least one employee to coordinate efforts to comply with Title II.[5] The public entity must also adopt grievance procedures and designate at least one employee who will be responsible for investigating any complaint filed under this grievance procedure.

---

[1] HR (ELC) at 84-85.
[2] Ibid.
[3] HR (ELC) at 85.
[4] H. Comm. on Educ. and Lab. Rep. at 98; S. Comm. on Lab. and Hum. Resources Rep. at 57-58.
[5] 28 C.F.R. section 35.107.

Title III — Public accommodations

Title III builds upon the foundation established by Congress under the Architectural Barriers Act and the Rehabilitation Act. Title III basically extends the prohibition against discrimination that existed for facilities constructed by or financed by the federal government to all private sector public facilities. Title III requires all goods, services, privileges, advantages, or facilities of any public place to be offered "in the most integrated setting appropriate to the needs of the [disabled] individual", except when the individual poses a direct threat to the safety or health of others. Title III additionally prohibits discrimination against individuals with disabilities in the full and equal enjoyment of all goods, services, facilities, etc.

Title III covers public transportation offered by private sector entities in addition to all places of public accommodation without regard to size. Congress wanted small businesses to have time to comply with this mandatory change without fear of civil action. To achieve this, Congress provided that no civil action could be brought against businesses that employ 25 or fewer employees and have annual gross receipts of $1 million or less between January 26, 1992 and July 26, 1992. Additionally, businesses with fewer than 10 employees and having gross annual receipts of $500,000 or less were provided a grace period from January 26, 1992 to January 26, 1993 to achieve compliance. Residential accommodations, religious organizations, and private clubs were made exempt from these requirements.

Title III provides categories and examples of places requiring public accommodations:

- Places of lodging, such as inns, hotels, and motels, except for those establishments located in the proprietor's residence and have not more than 5 rooms for rent
- Restaurants, bars, or other establishments serving food or drink
- Motion picture houses, theaters, concert halls, stadiums, or other place of exhibition or entertainment
- Bakeries, grocery stores, clothing stores, hardware stores, shopping centers, or other sales or rental establishments
- Laundromats, dry cleaners, banks, barber shops, beauty shops, travel services, funeral parlors, gas stations, offices of an accountant or lawyer, pharmacies, insurance offices, professional offices of health care providers, hospitals, or other service establishments
- Terminal depots, or other stations used for specified public transportation
- Parks, zoos, amusement parks, or other places of entertainment
- a nurseries, elementary, secondary, undergraduate schools, post-graduate private schools, or other places of education
- Daycare centers, senior citizen centers, homeless shelters, food banks, adoption agencys, or other social service center establishments
- Gymnasiums, health spas, bowling alleys, golf courses, or other places of exercise or recreation.[1]

---
[1] Title III section 310(7).

Managers should note that it is considered discriminatory under Title III for a covered entity to fail to remove structural, architectural, and communication barriers from existing facilities when the removal is readily achievable, easily accomplished, and can be performed with little difficulty or expense. Factors to be considered include the nature and cost of the modification, the size and type of the business, and the financial resources of the business among others. If the removal of a barrier is not readily achievable, the covered entity may make goods and services readily available and achievable through alternative methods to individuals with disabilities.

Managers should be aware that employers may not use application or other eligibility criteria that screen out individuals with disabilities unless they can prove that doing so is necessary to providing the goods or services that they provide to the public. Title III additionally makes discriminatory the failure to make reasonable accommodations in policies, business practices, and other procedures which afford access to and use of public accommodations to individuals with disabilities; it prohibits employers from denying access to goods and services because of the absence of auxiliary aids unless the providing of such auxiliary aids would fundamentally alter the nature of the goods or services or would impose an undue hardship. The ADA defines auxiliary aids and services as:

1. Qualified interpreters or other effective methods of making orally delivered materials available to individuals with hearing impairments,
2. Qualified readers, taped texts, or other effective methods of making visually delivered materials available to individuals with visual impairments,
3. Acquisition or modification of equipment or devices, and
4. Other similar services or actions.[1]

Title III does not specify the types of auxiliary aids that must be provided, but requires that individuals with disabilities be provided equal opportunity to obtain the same result as individuals without disabilities.

Title III provides an obligation to provide equal access, requires modification of policies and procedures to remove discriminatory effects, and provides an obligation to provide auxiliary aids in addition to other requirements. The safety and health exception and undue burden exception are available under Title III in addition to the structurally impracticable and possibly the disproportionate cost defenses for covered organizations.

Title IV — Telecommunications

Title IV amends Title II of the Communication Act of 1934[2] to mandate that telephone companies provide telecommunication relay services in their service areas by July 26, 1993. Telecommunication relay services provide individuals

[1] Title III section 3(1).
[2] 47 USC 201 **et. seq**.

with speech related disabilities the ability to communicate with hearing individuals through the use of telecommunication devices like the TDD systems or other non-voice transmission devices.

The purpose of Title IV is in large measure to establish a seamless interstate and intrastate relay system for the use of TDD's (telecommunication devices for the deaf) that will allow a communication-impaired caller to communicate with anyone who has a telephone, anywhere in the country.[1] Title IV contains provisions affording the disabled access to telephone and telecommunication services equal to that of a nondisabled community. In actuality, Title IV is not a new regulation but simply an effort to ensure that the general mandates of the Communication Act of 1934 are made effective. Title IV consists of two sections. Section 401 adds a new section (section 225) to the Communication Act of 1934 and section 402 amends section 711 of the Communications Act.

Regulations governing the implementation of Title IV were issued by the Federal Communication Commission (FCC) July 26, 1992. These regulations established the minimum standards, guidelines, and other requirements mandated under Title IV in addition to establishing regulations requiring round the clock relay service operations, operator maintained confidentiality of all messages, and rates for the use of the telecommunication relay systems which are equivalent to current voice communication services. Title IV prohibits the use of relay systems under certain circumstances, encourages the use of state-of-the-art technology where feasible, and requires public service announcements and other television programs which are partially or fully funded by the federal government to contain closed captioning.

Title V — Miscellaneous provisions

Title V is a myriad of provisions addressing a wide assortment of related coverage under the ADA. Title V permits insurance providers to continue to underwrite insurance, continue to use preexisting condition clauses, and to classify risks as long as consistent with state-enacted laws. Title V also permits insurance carriers to provide bona fide benefit plans based upon risk classifications but prohibits denial of health insurance coverage to an individual with a disability based solely on that person's disability.

Title V does not require special treatment in the area of health or other insurance for individuals with disabilities. An employer is permitted to offer insurance policies that limit coverage for a certain procedure or treatment even though this might have an adverse impact on the individual with a disability.[2] The employer or insurance provider may not, however, establish benefit plans as a subterfuge to evade the intent of the ADA.[3]

---

[1] H. Comm. on Energy and Com. Rep. at 28.
[2] H. Comm. on Educ. and Lab. Rep. at 59.
[3] EEOC Interpretive Guidelines, 56 Fed. Reg. 35,753 (July 26, 1991).

Title V provides that the ADA will not limit or invalidate other federal or state laws that provide equal or greater protection to individuals with disabilities. Additionally, the ADA does not preempt medical or safety standards established by federal law or regulation nor does it preempt state, county, or local public health laws. However, state and local governments and their agencies are subject to the provisions of the ADA, and courts may provide the same remedies (except punitive damages at this time) against state or local governments as any other public or private covered entity.

In an effort to minimize litigation under the ADA, Title V promotes the use of alternate dispute resolution procedures to resolve conflicts under the ADA. As stated in section 513, "Where appropriate and to the extent authorized by law, the use of alternate dispute resolution, including settlement negotiations, conciliation, fact-finding, mini-trials, and arbitration, is encouraged to resolve disputes under the ADA."[1] Safety and health professionals should note, however, the fact that the use of alternate dispute resolution is voluntary, and if used, the same remedies must be available as provided under the ADA.

Safety professionals should be knowledgeable regarding the ADA and provide appropriate consideration in the development of the emergency and disaster preparedness plan. Safety professionals can acquire a copy of the Americans With Disabilities Act as well as rules and interpretations provided by the EEOC, the Department of Labor, Centers for Disease Control, Department of Federal Contract Compliance as well as on the EEOC website at **www.eeoc.gov**.

Some of the common issues in the emergency and disaster preparedness program development for individuals with disabilities include the following:

- Individuals who are hearing impaired: alarm systems
- Individuals who are sight impaired: alarm systems and egress issues
- Individuals who are wheelchair bound: egress issues
- Individuals with psychological disabilities: training and egress issues
- Individuals with specific medical conditions: triage issues

Prudent safety professionals should remember that many of the usual methods of providing accommodation, such as elevators, may not be functioning in an emergency situation. Additionally, commonly used accommodations such as use of the *buddy system* can also create additional issues in the areas of absenteeism, turnover, and training.

In summation, prudent safety professionals should provide considerable thought and effort into the various issues of accommodation as part of the overall emergency and disaster preparedness planning process. However, safety professionals are cautioned to ensure that they are fully knowledgeable of the laws with regard to disability and ensure that they do not discriminate in any manner against employees with disabilities in the development or implementation of their emergency and disaster preparedness program.

---

[1] Title V section 513.

chapter twenty

Disaster preparedness assessments

> "The three great essentials to achieve anything worthwhile are, first, hard work; second, stick-to-itiveness; third, common sense."
>
> Thomas Alva Edison
>
> "Chaotic action is preferable to orderly inaction."
>
> Anonymous

An important component to ensure the quality and effectiveness of your emergency and disaster preparedness program is the periodic assessment and evaluation. In essence, each and every component of your overall program should be carefully analyzed individually, and in conjunction with other components, to ensure that the program element is operating efficiently and effectively. Additionally, this analysis and assessment should identify the strengths and weaknesses of the overall program efforts thus permitting the safety professional to focus his or her efforts to correcting the deficiencies in the upcoming time period.

Although a variety of methods can be utilized to appropriately assess your emergency and disaster program, the important factors are to ensure that the assessment instrument appropriately identifies all elements of the program and the assessment instrument appropriately and honestly provides an accurate assessment. A very simple method of providing an objective analysis is provided for your review and evaluation (see Example).

Remember the axiom *garbage in...garbage out*. Your assessment instrument should be carefully designed to acquire all of the necessary information to properly evaluate your program. Prudent safety professionals should strive to make the assessment as objective as possible and avoid subjective assessments where possible. Additionally, careful consideration should be provided to other motivators, such as monetary rewards, employment evaluations, and other factors, which may skew the information acquired during the assessment.

Example

	Disaster Preparedness	Answer		Total Points	Score
1.	Do you have a written disaster preparedness plan for your facility?	YES	NO	20	
2.	Do you have a written disaster preparedness responsibility list?	YES	NO	10	
3.	Do you have a written evacuation plan?	YES	NO	15	
4.	Are the evacuation routes posted?	YES	NO	10	
5.	Do you have emergency lighting?	YES	NO	10	
6.	Is the emergency lighting inspected on a weekly basis?	YES	NO	10	
7.	Do you have a written natural disaster plan? (AKA tornado, hurricane)	YES	NO	10	
8.	Is there a notification list for local fire department, ambulance, police, and hospital?	YES	NO	10	
9.	Are all supervisory personnel aware of their responsibilities in an evacuation?	YES	NO	10	
10.	Have triage and identification areas been designated?	YES	NO	5	
11.	Has an employee identification/notification procedure been developed for the evacuation procedure?	YES	NO	5	
12.	Have command posts been designated?	YES	NO	5	
13.	Do you have quarterly meetings to review with your disaster preparedness staff?	YES	NO	10	
14.	Do you have mock evacuation drills?	YES	NO	10	
15.	Do you possess a bomb threat procedure?	YES	NO	10	
16.	Is your management team in full understanding of the bomb threat procedure?	YES	NO	10	
	SECTION TOTAL			**160**	

	Medical	Answer		Total Points	Score
1.	Is your medical staff fully qualified?	YES	NO	10	
2.	Do you have all necessary equipment on hand?	YES	NO	10	
3.	Are you inventorying and purchasing necessary medical supplies in bulk in order to achieve the best price?	YES	NO	5	
4.	Are you equipped for a trauma situation? (i.e., air splints, oxygen, etc.)	YES	NO	15	
5.	Is your medical staff fully trained in the plant system?	YES	NO	10	
6.	Is your medical staff fully trained in the individual state requirements?	YES	NO	10	
7.	Does your dispensary have an emergency notification with the telephone numbers of the ambulance, hospital, etc?	YES	NO	10	

8.	Is your staff fully trained in the post offer screening and physical examination procedures?	YES	NO	10
9.	Does you medical staff have daily communication with the insurance and workers compensation administrators?	YES	NO	10
10.	Does your medical staff contact local physicians and hospitals on time loss claims?	YES	NO	10
11.	Does you medical staff track all injuries and lost time cases?	YES	NO	15
12.	Does your medical staff conduct the home and hospital visitation program?	YES	NO	5
13.	Are your trauma kits inspected, cleaned, and restocked on a weekly basis?	YES	NO	10
14.	Is your medical staff involved in local medical community activities?	YES	NO	5
15.	Does your medical staff tour the plant and know each area of the plant?	YES	NO	5
16.	Is your medical staff conducting yearly evaluations on employees using the respiratory equipment?	YES	NO	10
17.	Are the first aid boxes and stretchers inspected on a weekly basis?	YES	NO	10
18.	Is your medical staff fully trained in the proper use of the alcohol/controlled substance testing equipment?	YES	NO	20
19.	Is your medical staff conducting the alcohol/controlled substance testing properly?	YES	NO	20
	SECTION TOTAL			**195**

After completion of your emergency and disaster program assessment, it is vitally important to utilize the information which you have acquired to strengthen your program and correct any and all deficiencies identified. If the information is not utilized, safety professionals will be "spinning their wheels and wasting the effort" of the overall assessment. Remember, your emergency and disaster program must function properly at a moment's notice. Development of a program without periodic assessment will permit the program to become stale, and it will not function appropriately when needed. Periodic assessment is an absolute **must** and an important component of your overall program's success.

chapter twenty-one

Personal disasters — use of criminal sanctions

"Crime, like virtues, has its degrees."

Racine

"He who excuses himself accuses himself."
(Qui s'excuse, s'accuse.)

Anonymous

A relatively new phenomenon that is substantially impacting the scope of potential liability for safety and health professionals after a disaster situation is the use of state criminal laws by state and local prosecutors for injuries and fatalities which were incurred during the disaster and on the job. This new utilization of the standard state criminal laws in a workplace setting normally governed by OSHA or state plan programs is controversial, but appears to be a viable method through which states can penalize safety professionals and other corporate officials in situations involving fatalities or serious injury. The utilization of state criminal laws is not preempted by the OSHA Act and at least one state court has stated that issuance and payment of OSHA penalties does not bar the use of criminal sanctions under the theory of double jeopardy.[1]

It should be clarified that the use of criminal sanctions for workplace fatalities and injuries following a disaster situation is not a new area of concern. In Europe, criminal sanctions for workplace fatalities are frequently used. As far back as 1911, in the U.S., criminal sanctions were used in the well-known Triangle Shirt fire in New York in which over 100 young women were killed. The co-owners of the Triangle Shirt Company were indicted on criminal manslaughter charges although subsequently acquitted of these charges. Safety and health professionals should, however, take note that the sources of the potential criminal liability (i.e., state criminal codes in addition to the OSHA Act) and the enforcement frequency (i.e., increased utilization of criminal charges under the OSHA Act and state criminal codes) is a recent trend.

[1] See recent decision in Commonwealth of Kentucky.

Utilizing the individual state's criminal code, state and/or local prosecutors are taking an active role in workplace safety and health through the enforcement of state criminal sanctions against employers for on-the-job deaths and serious injuries. The area of workplace safety and health has, since the enactment of the OSHA Act in 1970, been exclusively within the domain of OSHA and state plan programs. State prosecutors have recently challenged OSHA's federal preemption of this area and have created an entirely new area of potential personal criminal liability that the safety and health professional should be prepared to address on an individual and corporate basis in the event of a disaster situation.

The case that highlighted the issue of whether OSHA has jurisdiction over workplace injuries and fatalities, and thus *preempts* state prosecution under state criminal statutes for workplace deaths, involved the first degree murder convictions by an Illinois court of the former president, the plant manager, and the plant foreman of Film Recovery Systems, Inc., for the 1983 work-related death of an employee.[1] In this case, Stephan Golab, a 59-year-old immigrant employee from Poland, died as a direct result of his work at the Elk Grove, IL, manufacturer in which he stirred tanks of sodium cyanide used in the recovery of silver from photographic films. In February 1983, Mr. Golab walked into the plant's lunchroom, started violently shaking, collapsed, and died from inhaling the plant's cyanide fumes.[2] Following his death, both OSHA and the Cook County state's attorney office investigated the accident. OSHA found 20 violations and fined the corporation $4850. This monetary penalty was later reduced by half.[3] The Cook County prosecutor's office, on the other hand, took a different view and filed charges of first degree murder and 21 counts of reckless conduct against the corporate officers and management personnel, and issued involuntary manslaughter charges against the corporation itself.[4]

[1] *People v. O'Neil (Film Recovery Systems)*, Nos. 83 C 11091 & 84 C 5064 (Cir. Ct. of Cook County, IL. June 14, 1985), **rev'd**, 194 Ill. App.3d 79, 550 N.E.2d 1090 (1990).
[2] Gibson, "A Worker's Death Spurs Murder Trial," *The National Law Journal*, January 21, 1985 at 10.
[3] Note: *Getting Away With Murder - Federal OSHA Preemption Of State Criminal Prosecutions for Industrial Accidents*, 101 Harv L.J. 220 (1987). The violations ranged from failure to instruct employees about the hazards of cyanide and to provide first-aid kits with antidotes for cyanide poisoning to failure to keep the floors clean and dry. OSHA did not cite Film Recovery Systems, Inc. for exceeding the permissible exposure limit for cyanide. See Citations and Notification of Penalty issued by OSHA to Film Recovery Systems, Inc., Report #176 (Mar. 11, 1983) and Amendment (Mar. 30, 1983).
[4] Ibid. *Also see*, Brief for Appellee at 12-20; [The state presented evidence that the deceased employee and many of the other Film Recovery Systems' employees who worked around unventilated tanks often experienced headaches, nausea, burning eyes, and burning skin from cyanide fumes. The plant management never informed the employees that they were working with a deadly toxin and provided them with virtually no safety equipment. Most of the workers were illegal aliens, primarily from Mexico and Poland, and thus could not read the warnings and were vulnerable to losing their jobs if they complained] and Gibson, "A Worker's Death Spurs Murder Trial," *The National Law Journal*, Jan. 21, 1985 at 10; [Ms. Erpito (secretary at Film Recovery Systems, Inc.) said she was instructed by her bosses never to use the word cyanide around workers. She testified that when going into the plant, "your eyes started to burn and you got a headache."]

Under Illinois law, murder charges can be brought when someone "knowingly creates a strong probability of death or great bodily harm" even if there is no specific intent to kill.[1] The prosecutor's office initially brought charges against five officers and managers of the corporation, but the defendant successfully fought extradition from another state.[2] The prosecutor's intent was to *criminally pierce the corporate veil*, to place liability not only upon the corporation but on the responsible individuals.[3]

During the course of the trial, the Cook County prosecutor presented extensive and overwhelming testimony that the company officials knew of unsafe conditions, knowingly neglected the unsafe conditions, and attempted to conceal the dangers from employees. Witnesses testified to the following:

- Company officials exclusively hired foreign workers who were not likely to complain to inspectors about working conditions and would perform work in the more dangerous areas of the plant.
- Officials instructed support staff never to use the word cyanide around the workers.
- According to the testimony of two co-workers, Mr. Golab complained to plant officials and requested to be moved to an area where the fumes were not as strong shortly before his death. These pleas were ignored.
- Testimony of numerous employees referred to recurrent nausea, headaches, and other illnesses.
- Employees were not issued adequate safety equipment or warned of the potential dangers in the workplace.
- An industrial saleswoman reported trying unsuccessfully to sell safety equipment to the owners.
- The supervisors instructed employees to paint over the skull-and-crossbones on the steel containers of cyanide-tainted sludge and to hide the containers from inspectors after the employee's death.
- Workers were never told they were working with cyanide and were never told of the hazardous nature of this substance.
- Employees were grossly overexposed on a daily basis and, after the incident, the company installed emission-control devices that dropped cyanide emissions twenty-fold.[4]

---

[1] Moberg, David, et al., "Employers Who Create Hazardous Workplaces Could Face More Than Just Regulatory Fines, They Could Be Charged With Murder," 14 *Student Lawyer*, Feb. 1986 at 36. (The legislative intent may have had cases like arsonist torching in mind but the principle is easily extended to cases such as *Film Recovery Systems, Inc*.)

[2] Gibson, "A Worker's Death Spurs Murder Trial," *The Nat Law J.*, Jan. 21, 1985 at 10. (The original indictment before the grand jury in October 1983 was against Steven J. O'Neill, the former president; Michael McKay, an officer; Gerald Pett, vice-president; Charles Kirschbaum, plant manager; and Daniel Rodriguez, a plant foreman. Mr. McKay successfully fought extradition from Utah in February 1984.)

[3] Moberg, David, et al., "Employers Who Create Hazardous Workplaces Could Face More Than Just Regulatory Fines, They Could Be Charged With Murder," *supra* at 36.

[4] *People v. O'Neil, et. al. (Film Recovery Systems)*, Nos. 83 C 11091 & 84 C 5064 (Cir. Ct. of Cook County, Ill. June 14, 1985), **rev'd**, 194 Ill. App.3d 79, 550 N.E.2d 1090 (1990).

Given the extreme circumstances in this case, the prosecution was able to obtain conviction of three corporate officials for murder and 14 counts of reckless conduct; each was sentenced to 25 years in prison and the corporation was convicted of manslaughter and reckless conduct and fined $24,000.[1] The court rejected outright the company's defense of preemption of the state prosecution by the federal OSHA Act. This case opened a new era in industrial safety, and as the prosecutor appropriately stated:

> These verdicts mean that employers who knowingly expose their workers to dangerous conditions leading to injury or even death can be held criminally responsible for the results of their actions. Today's [criminal] verdict should send a message to employers and employees alike that the criminal justice system can and will step in to protect the rights of every worker to a safe environment and to be informed of any hazard that might exist in the work place.[2]

As expected, the decision in *People v O'Neil* opened a new era in workplace health and safety.[3] The case marked the first time in which a corporate officer had been convicted of murder in a workplace death. As expected, once the door was opened, prosecutors across the country began to initiate similar action, such as in the Imperial Foods fire in which the plant owner plead guilty to manslaughter and received a 20-year prison sentence,[4] and institute programs, such as the Los Angeles County District Attorney's *roll out* program,[5] to address work place accidents.[6]

The significance of the Film Recovery case, according to Professor Ronald Jay Allen of Northwestern University School of Law, "is more psychological than anything else. It will sensitize prosecutors to the possibility of bringing criminal charges against the officials of a company when an egre-

---

[1] Sand, Robert, "Murder Convictions For Employee Deaths; General Standards Versus Specific Standards," 11 *Employee Rel J.* 526 (1985-86).
[2] Ranii, David, "Verdict May Spur Industrial Probes," *Nat L. J.*, July 1, 1985 at 3.
[3] *People v. O'Neil, et. al. (Film Recovery Systems)*, Nos. 83 C 11091 & 84 C 5064 (Cir. Ct. of Cook County, IL. June 14, 1985), **rev'd**, 194 Ill. App.3d 79, 550 N.E.2d 1090 (1990).
[4] Jefferson, "Dying for Work," *ABA Journal*, Jan. 1993 at 48.
[5] Ibid.
[6] Middleton, Martha, "Get Tough On Safety," *Nat L. J.*, April 21, 1986 at p. 1. [Note: Austin, Texas, prosecutors filed criminal charges in two cases in which workers were killed in trench cave-ins; Los Angeles district attorney Ira K. Reiner ordered a new *roll out* program in which an attorney and an investigator would be sent to the scene of every industrial workplace death; Milwaukee County District Attorney E. Michael McCann ordered investigators to begin checking every workplace death for possible criminal violations; U.S. Secretary of Labor William E. Brock referred the case of Union Carbide Corporation (pesticide violations in Institute, WV facility) to the Justice Department. *Also see* Sand, Robert, "Murder Convictions For Employee Deaths: General Standards Versus Specific Standards," 11 *Employee Rel. J.* 526 (1985-86) [Cook County prosecutors initiated another action immediately after this case against five executives of a subsidiary of North American Philips (Chicago Wire)].

gious accident takes place."[1] This new attitude toward work-related deaths was summarized by Los Angeles prosecutor John Lynch when he stated, "We [prosecutors] have to raise the consciousness that it is possible to have a criminal homicide in a case where there is no gun."[2] Prosecutors argue that this type of criminal prosecution is needed because of the lax regulatory enforcement of workplace safety by OSHA, the duty of the local prosecutors to the employees, and a need to send a message to employers that they have a responsibility for workplace safety. On the other hand, defense lawyers argue liability for workplace accidents and deaths should continue to be handled in the regulatory and civil arenas and should not intrude into the criminal area.

Following the decision in *People v. O'Neil*, several criminal prosecutions were initiated in other states involving work-related deaths. Prosecutors acknowledge that the preemption defense in the early cases was a major obstacle in gaining convictions against employers. As explained by Jay C. Magnuson, deputy chief of the Public Interest Bureau of the Cook County State's Attorney's office who presented the Illinois cases, "If you can't even charge someone, you're pretty much out of the ball game."[3]

Prosecutors across the country have successfully defeated a number of defenses, including preemption, and are becoming more creative in filing charges and handling workplace injury and fatality cases. Two of the areas that generated a substantial amount of litigation following *People v. O'Neil* were whether preemption attached to the general OSHA standards only (including the general duty clause) or to the specific OSHA standards only and whether a specific OSHA standard preempts a more general industry standard even though the specific hazard falls between the cracks of coverage under the specific standard.[4]

The major question to be addressed by the courts after the Film Recovery case was whether the OSHA Act in total preempts any or all state criminal actions. With *People v. O'Neil* and several other similar convictions on appeal at that time,[5] the interest turned to the Illinois Appeals court decision in *People v. Chicago Magnet and Wire Corporation*.[6]

In this case, the Grand Jury of Cook County indictments against the corporation and five corporate officers charging each with multiple counts of aggravated battery, reckless conduct, and conspiracy pursuant to the

---

[1] Ranii, David, "Verdict May Spur Industrial Probes," *Nat L. J.*, July 1, 1985 at 7.
[2] Middleton, Martha, "New Worry For Companies," *Nat L. J.*, April 21, 1986 at p. 9.
[3] Middleton, Martha, "Death In The Workplace: A Crime?," *Nat L. J.*, July 13, 1987 at p. 3.
[4] Sand, Robert, "Murder Convictions For Employee Deaths; General Standards Versus Specific Standards," 11 *Employee Rel J.* 526 (1985-86).
[5] *See* N 1; *Also see Sabine Consolidated*, no. 3-87-051 CR (Ct. App. TX. 1987); *People v. Pymm*, NY Times, Nov. 14, 1987, at 31, col 1 (NY Sup. Ct. Nov. 13, 1987) (A trial court in New York held that the OSHA Act preempted a state criminal prosecution of the owners of a thermometer plant who were charged with exposing workers to unsafe levels of mercury fumes. The jury found the defendants guilty of assault, reckless endangerment, conspiracy, and falsifying business records, but the judge set aside the verdict).
[6] 157 Ill App. 3d 797, 510 N.E. 2d 1173 (1987).

Illinois criminal code. The trial court dismissed all charges against the employer finding that OSHA preempted Illinois from applying Illinois criminal law to conduct involving federally regulated occupational safety and health issues within the workplace.[1]

On appeal, the State contended that the indictments should be reinstated because the police power of the state was neither expressly or indirectly preempted by Congress under the Occupational Safety and Health Act. The State argued, relying on *Silkwood v. Kerr-McGee Corporation*,[2] that state laws are only preempted when Congress has declared its intent to occupy a certain area of the law, and, where preemptive intent is not expressly stated in the statute, the intent of Congress must be derived from the statutory language, the comprehensiveness of the regulatory scheme, the legislative history, and the specific conflict between the state and federal statutes in question.[3]

The State additionally argued that the prosecution was not preempted because it was based on the application of general criminal law rather than on the enforcement of specific workplace health and safety regulations. To support its assertion that Congress explicitly intended to leave preexisting state criminal laws undisturbed, the State cited section 553(b)(4) of the Act which provides:

> Nothing in this chapter shall be construed to supersede or in any manner affect any workers' compensation law or to enlarge or diminish or affect in any other manner the common law or statutory rights, duties, or liabilities of employers and employees under any law with respect to injuries, diseases, or death of employees arising out of, or in the course of employment. 29 U.S.C. section 553(b)(4) (1982).[4]

The Illinois Court of Appeals rejected the *Silkwood*[5] preemption approach holding that the Atomic Energy Act, relied upon in *Silkwood*,[6] was not applicable to situations governed by the OSHA Act. Additionally, the court rejected the state's position on section 553(b)(4) of the Occupational Safety and Health Act stating, "the state would not be foreclosed from applying its criminal laws in the workplace if the prosecution charged the defendants with crimes not involving working conditions." The Court also relied on the fact that the State of Illinois had the opportunity to retain responsibility for safety and health in the workplace by initiating a *state plan* occupational

---

[1] *People v. Chicago Magnet Wire Corporation*, 126 Ill. 2d 356, 129 Ill. Dec. 517, 57 U.S.L.W. 2460, 534 N.E. 2d 962 (1989).
[2] 454 US. 239, 104 S. Ct. 519, 78 L. Ed. 2d 443 (1984); *Also see Jones v. Rath Packing Co.*, 430 U.S. 519, 97 S. Ct. 1305, 51 L. Ed. 2d 504 (1977).
[3] *People v. Chicago Magnet Wire Corp., supra. Also see, Hillsborough County, Florida v. Automated Medical Laboratories*, 471 U.S. 707, 105 S. Ct. 2371, 85 L.Ed. 2d 714 (1985).
[4] Ibid.
[5] 454 US. 239, 104 S. Ct. 519, 78 L. Ed. 2d 443 (1984).
[6] Ibid.

Chapter twenty one: Personal disasters — use of criminal sanctions 161

safety and health program that would have preempted the federal Occupational Safety and Health Act, but although a state plan was submitted, this plan was withdrawn by the State.[1]

In affirming the decision of the circuit court, the appeals court stated, "The State has expressed valid and legitimate concerns about the consequences of preemption on its ability to control the activities of employers. But Congress has evidenced an intent that criminal sanctions should not be imposed for activities involving workplace health and safety except in highly limited circumstances, and that health and safety requirements should be established through standard setting, which provides employers with clear and detailed notice of their legal obligations. Illinois' view that employers may be held criminally liable for workplace injuries and illnesses, regardless of their compliance with OSHA standards, would lead to piecemeal and inconsistent prosecutions of regulatory violations throughout the states, a result that Congress sought to preclude in enacting OSHA."[2]

In the Chicago Magnet and Wire case, the trial court dismissed the indictments against the manufacturer and the corporate officers finding that the Occupational Safety and Health Act preempted the state of Illinois prosecution. The circuit court affirmed this decision finding the Act preempts the state prosecution unless the state had received approval from OSHA officials to administer its own occupational safety and health plan.[3]

At the circuit court level, the state relied extensively yet unsuccessfully on the Supreme Court's decision in *Silkwood v. Kerr-McGee Corporation* which held that even though the federal government occupies the field of nuclear safety regulations, the state courts were not preempted from assessing punitive damages for work-related radiation exposure.[4] The circuit court explicitly rejected the analogy to *Silkwood* on two grounds: first, that *Silkwood* was decided under the Atomic Energy Act rather than the Occupational Safety and Health Act; and second, the criminal laws, unlike punitive damages, are meant to regulate conduct rather than compensate victims.[5]

Additionally, the circuit court rejected the state's arguments of waiver of authority and inadequacy of OSHA, and held that section 18 should be interpreted narrowly based upon the comprehensiveness and legislative history of the OSHA Act.[6]

---

[1] Ibid. See also, *Stanislawski v. Ind. Comm.*, 99 Ill. 2d 36, 75 Ill. Dec. 405, 457 N.E. 2d 399 (1983) (State plans); *United Airlines v. Occ. Safety and Health Appeal Board*, 32 Cal. 3d 762, 187 Ca. Rptr. 397, 654 P. 2d 157 (1982) (a state is preempted from regulating matters governed by OSHA standards in the absence of the adoption of a federally approved state plan); *Five Migrant Farmworkers v. Hoffman*, 135 N.J. Super. 242, 345 A. 2d 370 (1975) (OSHA supersedes all state laws with respect to general working conditions).
[2] Ibid.
[3] *Chicago Magnet, supra* at p 10.
[4] 464 US. 238 (1984); (The *Silkwood* court ruled punitive damages were not preempted, even though they were regulatory and were not purely compensatory, and found that punitive damages serve the purpose of punishing the employer and deters future misconduct.)
[5] *Chicago Magnet, supra*
[6] Ibid.

On February 2, 1989, the Supreme Court of Illinois addressed the issue "whether the OSHA Act of 1970 (OSHA) [29 U.S.C. section 651 et seq. (1982)] preempts the state from prosecuting the defendants, in the absence of approval from OSHA officials, for conduct which is regulated by OSHA occupational health and safety standards."[1] In a landmark decision, the court reversed and held that OSHA Act did not preempt the state from prosecuting the defendants.[2]

The court initially addressed the extent to which state law was preempted by federal legislation under the Supremacy Clause of the Constitution. Preemption "is essentially a question of congressional intentions" and "thus, if Congress, when acting within constitutional limits, explicitly mandates the preemption of state law... we [court] need not proceed beyond the statutory language to determine that state law is preempted."[3] "Even absent an express command by Congress to preempt state law in a particular area, preemptive intent may be inferred where the scheme of federal regulation is sufficiently comprehensive to make reasonable the inference that Congress left no room for supplementary state regulation."[4] The Court also noted that preemptive intent could also be inferred "where the regulated field is one is which the federal interest is so dominant that the federal system will be assumed to preclude enforcement" or "where the object sought to be obtained by the federal law and the character of obligations imposed by it ...reveal the same purposes."[5]

The defendant initially argued that Congress, in section 18(a) of the Act, explicitly provided that the states are preempted from asserting jurisdiction unless approval for a state plan was acquired from OSHA officials under section 18(b).[6] (section 18 provides: (a) Nothing in this chapter shall prevent any state agency or court from asserting jurisdiction under state law over any occupational safety or health issue with respect to which no standard is in effect under section 655 of this title; (b) Any state which, at any time, desires to assume responsibility for development and enforcement therein of occupational safety and health standards relating to any occupational safety or health issue with respect to which a federal standard has been promulgated under section 665 of this title shall submit a state plan for the

---

[1] Ibid. at 6.
[2] Ibid.
[3] Ibid. at p. 10; *Also see Malone v. White Motor Corporation* (1978), 435 U.S. 497, 504, 98 S. Ct. 1185, 1190, 55 L. Ed. 2d 443, 450; *Retail Clerks International Association, Local 1625 v. Schermerhorn* (1963), 375 U.S. 96, 84, S. Ct. 219, 11 L.Ed. 2d. 179; *Pacific Gas & Electric Co. v. State Energy Resources Conservation & Development Comm.*, (1983), 461 U.S. 190, 302, 103 S.Ct. 1713, 1722, 75 L.Ed. 2d 752, 765.)
[4] Ibid. at 10; *Also see Hillsborough County v. Automated Medical Laboratories, Inc.*, (1985) 471 U.S. 707, 713, 105 S.Ct. 2371, 2375, 85 L.Ed.2d 714, 721; *Rice v. Sante Fe Elevator Corporation*, (1947), 331 U.S. 218, 230, 67 S.Ct. 1146, 1152, 91 L.Ed. 1447, 1959.
[5] Ibid. *Also see, Rice v. Sante Fe*, supra at 230; *Hines v. Davidowitz* (1941), 312 U.S. 52, 61 S.Ct. 399, 85 L.Ed. 581; *Fidelity Federal Savings & Loan Assoc. v. de la Cuesta* (1982), 458 U.S. 141, 153, 102 S.Ct. 3014, 3022, 73 L.Ed.2d 664, 675.
[6] Ibid. at 13.

Chapter twenty one: Personal disasters — use of criminal sanctions

development of such standards and their enforcement. 29 U.S.C. section 667 (1982)). Defendant argued that the narrow interpretation of section 18 by the lower courts is consistent with the legislative history of the Act.

Notwithstanding, section 18 was interpreted by the court to invite state administration of their own safety and health plans and was not intended to preclude supplementary state regulation.[1] Section 2 of the Act provided that states are to assume the fullest responsibility for the administration and enforcement of their safety and health laws.[2] The Court additionally examined the legislative history of the Act noting that "it is highly unlikely that Congress considered the interaction of OSHA regulations with other common law and statutory schemes other than workers' compensation" and "it is totally unreasonable to conclude Congress intended that OSHA's penalties would be the only sanctions available for wrongful conduct which threatens or results in serious physical injury or death to workers."[3]

The court found that Congress sought to develop uniform national safety and health standards and the purpose underlying section 18 was to ensure that OSHA would create a nationwide floor of effective safety and health standards and provide for the enforcement of those standards.[4] Additionally, while additional sanctions imposed through state criminal law enforcement for conduct also governed by OSHA safety standards may incidentally serve as a regulation for workplace safety, there is nothing in OSHA or its legislative history to indicate that because of its incidental regulatory effect.[5] The Court concluded that "it seems clear that the federal interest in occupational health and safety...(is)...not to be exclusive."[6]

The Supreme Court, unlike the lower courts, viewed the decision in *Silkwood v. Kerr-McGee Corporation*[7] as applicable to this preemptive issue under the Act. The lower courts explicitly rejected the analogy to *Silkwood* on the grounds that *Silkwood* was decided under the Atomic Energy Act rather than OSHA and criminal laws, and unlike punitive damages, were meant to regulate conduct rather than compensate victims. The Supreme Court noted that the *Silkwood* court addressed "a question with resemblance to the one here" and there is little if any difference in the regulatory effect of punitive damages in tort and criminal penalties under the criminal law.[8] The Court reversed holding "we find no reason...why what the Court declared in *Silkwood* should not be applied to the preemptive effect of OSHA."[9] Additionally, the Court noted, "if Congress, in OSHA, explicitly declared it was willing to accept the incidental regulation imposed by

---

[1] Ibid. at 17.
[2] 29 U.S.C. section 651 (1982).
[3] *Chicago Magnet, supra*
[4] Ibid.; *Also see United Airlines, Inc. v. Occupational Safety and Health Appeals Board* (1982), 32 Cal. 3d. 762, 654 P. 2d 157, 187 Cal. Rptr. 387.
[5] Ibid. at 20.
[6] Ibid.
[7] 464 US. 238, 104 S. Ct. 615, 78 L.Ed. 2d 443 (1984).
[8] Ibid.
[9] Ibid.

compensatory damages awards under state tort law, it cannot plausibly be argued that is also intended to preempt state criminal law because of its incidental regulatory effect on workplace safety."[1]

The Court was similarly unconvinced by the defendant's contention that "it is irrelevant that the State is invoking criminal law jurisdiction as long as the conduct charged is an indictment or information is conduct subject to regulation by OSHA."[2] Defendant additionally argued "that the test of pre-emption is whether the conduct ...is in any way regulated by OSHA...and the conduct charged in the indictment is conduct regulated by OSHA."[3]

The Court rejected this argument finding that "simply because the conduct sought to be regulated in a sense under state criminal law is identical to that conduct made subject to federal regulation does not result in state law being preempted. When there is no intent shown on the part of Congress to preempt the operation of state law, the inquiry is whether there exists an irreconcilable conflict between the federal and state regulatory schemes."[4] The Court noted that a conflict exists only when "compliance with both federal and state regulations is a physical impossibility" or when state law "stands as an obstacle to the accomplishment and execution of the full purposes and objectives of Congress."[5] The Court could find no existing conflict or obstacle between OSHA and state's criminal law which would prohibit state's enforcement of this criminal action.

The defendant argued that state criminal prosecution would conflict with the purposes under OSHA in that Congress intended that the federal government was to have exclusive authority to set occupational safety and health standards.[6] "The standards were to be set only after extensive research to assure that the standards would minimize injuries in the workplace but at the same time not be so stringent that compliance would not be economically feasible."[7] Although the Court rejected this argument, it was noted that the defendant correctly pointed out that although states are given an opportunity to enforce their own safety and health standards under a state plan which are at least as effective as the federal program, OSHA retains jurisdiction until the state plan is approved.[8]

The Court also rejected the defendant's closely related argument that "federal supervision over state efforts to enforce their own workplace health

---

[1] Ibid.
[2] Ibid.
[3] Ibid.
[4] Ibid. at 23; *Also see Rice v. Norman Williams Co.* (1982), 458 U.S. 654, 659, 102 S.Ct. 3294, 3298-99, 73 L.Ed.2d 1042, 1049; *Huron Protland Cement Co. v. City of Detroit* (1960), 362 U.S. 440, 443, 80 S.Ct. 813, 815, 4 L.Ed.2d 852, 856; *Amal. Assoc. of Street, El. Ry. & Motor Coach Employees of Am. v. Lockridge* (1971), 403 U.S. 274, 285-86, 91 S.Ct. 1909, 1917, 29 L.Ed.2d 473, 482.
[5] Ibid. at p. 23; *Also see Florida Lime & Avocado Growers, Inc. v. Paul* (1963), 373 U.S. 132, 142-43, 83 S.Ct. 1210, 1217, 10 L.Ed.2d 248, 257; *Hines v. Davidowitz* (1941), 312 U.S. 52, 67, 61 S.Ct. 399, 404, 85 L.Ed. 581, 587.
[6] Ibid. at p. 24.
[7] Ibid.
[8] Ibid.

and safety programs would be thwarted if the state, with prior approval from OSHA officials, could enforce its criminal laws...[and] impose(s) standards so burdensome as to exceed the bounds of feasibility or so vague as not to provide clear guidance to employers."[1] The Court, in rejecting this argument, noted there was no finding that "state prosecutions of employers for conduct which is regulated by OSHA standards would conflict with the administration of OSHA or be at odds with its standards goals or purposes."[2] On the contrary, "prosecutions of employers who violate state criminal laws by failing to maintain safe working conditions for their employees will surely further OSHA's stated goal of assuring so far as possible every working man and woman in the nation safe and healthful working conditions."[3] The Court went further in stating, "state criminal law can provide valuable and forceful supplement to ensure that workers are more adequately protected and that particularly egregious conduct receives appropriate punishment."[4]

Along the same lines, the defendant argued the state did not possess the ability or resources to enforce more stringent safety and health standards than OSHA. The Court rejected this argument noting that the defendant's interpretation would "convert the statute...into a grant of immunity for employers responsible for serious injuries or deaths of employees." The Court further noted this would be a "consequence unforeseen by Congress" and "enforcement of state criminal law in the workplace will not stand as an obstacle to the accomplishment and execution of the full purposes and objectives of Congress."[5]

The Court noted that the preemption issue in the instant case has been addressed by very few courts. The appellate courts of the state of Michigan and Texas held that OSHA preempted state prosecutions[6] while the appellate court of Wisconsin in *State ex rel. Cornellier v. Black* held to the contrary.[7] In *Cornellier*, the officer/director of a fireworks manufacturer was charged with homicide by reckless conduct for his conduct regarding prior knowledge and disregard of safety violations which resulted in the death of an employee.[8] In finding that OSHA did not bar the prosecution and the complaint was sufficient to state probable cause, the court stated, "There is nothing in OSHA which we believe indicates a compelling congressional direction that Wisconsin, or any other state, may not enforce its homicide laws in the workplace. Nor do we see any conflict between the act and [the Wisconsin statute]. To the contrary, compliance with federal safety and health regulations is consistent, we believe, with the discharge of the state's duty

[1] Ibid.
[2] Ibid.
[3] Ibid. at 24-25; *Also see* 29 U.S.C. section 651(b) (1982).
[4] Ibid. at 25.
[5] Ibid.
[6] *People v. Hegedus* (1988), 169 Mich. App. 62, 425 N.W. 2d 729, leave to appeal granted in part (1988), 431 Mich. 870, 429 N.W. 2d 593; *Sabine Consolidated, Inc. v. State* (Tex. App. 1988), 756 S.W. 2d 865.
[7] *Cornellier v. Black* (Wisc. App. 1988), 144 Wisc. 2d 745, 425 N.W.2d 21.
[8] Ibid.; (Defendant's petition for a writ of habeas corpus was denied.)

to protect the lives of employees, and all other citizens, through penalty for violation of any safety regulations. It is only attempting to impose the sanctions of the criminal code upon one who allegedly caused the death of another person by reckless conduct. And the fact that ...conduct may in some respects violate OSHA safety regulations does not abridge the state's historic power to prosecute crimes."[1] The *Chicago Magnet Wire* court gave great deference to this decision by the Wisconsin court.

Subsequent to the appellate court's decision in *Chicago Magnet Wire*, the congressional committee on government operations issues a report and the Department of Justice's letter responded to the committee's report which directly addressed the issue in the instant case. On September 27, 1988, this congressional committee approved and adopted a report addressing the preemption issue.[2] In this report, the committee concluded that "inadequate use has been made of the criminal penalty provision of the Act and recommended to Congress that OSHA should take the position that the states have clear authority under the Federal OSHA Act, as it is written, to prosecute employers for acts against their employees which constitute crimes under state law."[3]

To supplement the congressional finding was a letter from the Department of Justice addressed to the committee which directly responded to the report. This letter, in part, stated that the Department of Justice shares the concern of the committee as to the adequacy of the statutory criminal penalties provided for violations of OSHA. This letter also observed, "As for the narrower issue as to whether the criminal penalty provisions of the OSHA Act were intended to preempt criminal law enforcement in the workplace and preclude the states from enforcing against the employers the criminal laws of general application, such as murder, manslaughter, and assault, it is our view that no such general preemption was intended by Congress. As a general matter, we see nothing in the OSHA Act or its legislative history which indicates that Congress intended for the relatively limited criminal penalties provided by the Act to deprive employees of the protection provided by state criminal laws of general applicability."[4]

The defendants argued that the Congressional Committee's findings and the Justice Department's letter were not binding on the Court and urged the Court to provide little deference to these reports. The Court agreed that these reports were not binding but noted "it is certainly not inappropriate to note ... the view of the governmental department charged with the enforcement of OSHA..."[5]

[1] Ibid.
[2] Report of House Committee on Government Operations, *Getting Away With Murder In The Workplace: OSHA's Non-use of Criminal Penalties for Safety Violations*, HR. Rep. No. 1051, 100th Cong., 2d Sess. 9 (1988).
[3] Ibid.
[4] Letter of Justice Department to House Committee on Government Operations (1988).
[5] *Chicago Magnet, supra* at 28.

Given the conclusions reached by the Congressional Committee and the Justice Department subsequent the appellate courts review and the arguments addressed above, the Court held "that the state… [was] not preempted from conducting prosecutions" by the OSHA Act against employers who consciously expose employees to hazards in the workplace.[1]

As expected, this decision by the Illinois Supreme Court has been appealed to the U.S. Supreme Court and thus this issue is not completely resolved as of this writing. However, since the *Chicago Magnet Wire* decision, two other state supreme courts have similarly rejected preemption arguments.[2] The current decisions of the state supreme courts in rejecting the preemption issue set a precedent for other states to test this methodology and begin to utilize the state criminal codes in situations involving workplace accidents. A possible decision by the U.S. Supreme Court could settle the preemption issue, but no decision is pending at the time of this writing. The use of state criminal code enforcement in incidents involving workplace injuries and fatalities by state prosecutors has greatly expanded the potential liability to corporate officers beyond that of the OSHA Act who willfully neglect their safety and health duties and responsibilities to their employees. This increased potential liability for all levels of the management hierarchy may affect the structure and methodology utilized for the management of health and safety in the American workplace in the future.

Safety and health professionals should be aware that individual states can create special legislation, such as Maine's Workplace-Manslaughter law of 1989, which governs workplace injuries and fatalities within the state. This type of specialized state law is applicable only in the state in which the legislation was enacted and is normally enforceable under the powers of the individual state. This type of state-enacted law usually requires compliance with the state law in addition to the federal OSHA standards or state plan requirements. Safety and health professionals should also be aware that state legislatures may modify existing laws, such as California's SB-198 that modifies the workers' compensation laws, which can significantly change the responsibilities in the safety and health area and possible civil and criminal sanctions for noncompliance.

Safety and health professionals facing a fatality, serious injury, or multiple injury type situation after a disaster situation should be prepared for an investigation by the state or local prosecutor's office, state police, or other investigative law enforcement agency. Although the odds are in favor that the criminal investigation will go no further than the investigation stage unless the situation involves willful, reckless, or abnormally negligent conduct on the part of the safety and health professional and company, safety and health professionals should be prepared for a substantial and probing

---

[1] Ibid.
[2] *People v. Pymm*, 76 N.Y.2d 511, 563 N.E. 2d 1 (1990); *People v. Hegedus*, 432 Mich. 598, 443 N.W. 2d 127 (1989).

inquiry. Safety and health professionals should be aware that the prosecutor's investigation will be made from a criminal prospective rather than an OSHA prospective and thus all rights should be preserved until such time as legal counsel can be consulted.

In most criminal investigations, it is advisable to error to the conservative. As is depicted on various police-related television programs, one is provided specific rights guaranteed under the U.S. Constitution (i.e., you have the right to remain silent, you have a right to an attorney...). Simply because the fatality or injury occurred at the worksite does not mean that the safety and health professional or other corporate official has waived these rights.

In most circumstances, the prosecutor or detective investigating the accident will not read the *Miranda* rights to the company representative until an arrest is to be made of the individual. However, safety and health professionals should be aware that most comments, photographs, and other evidence gathered during the investigation can be used in a court of law. Additionally, following an arrest and the reading of the rights, an individual possesses the ability to waive his or her rights through a verbal or written acknowledgment. Any comments made following the waiving of the *Miranda* rights can be used as evidence.

Normally, in situations involving workplace fatalities or injuries, the investigation is concluded without immediate arrests. The prosecutor will evaluate the evidence, and if substantial, normally submits the evidence to a grand jury. If the grand jury finds the evidence sufficiently substantial, than an indictment is rendered and arrest warrants are issued for the individuals in question.

Following arrest, there is normally a preliminary hearing where the charges are read to the individual and they are asked to plead. If the individual cannot afford legal counsel, counsel is normally appointed at that time. A hearing date is normally scheduled at this time and bond is usually set.

Safety and health professionals should be prepared for a varying degree of isolation from the other named corporate officials and possibly from the company following arrest. Depending on the circumstances, the safety and health professional may be required to acquire his or her own legal counsel and bear the burden of this cost. Abandonment or isolation of the safety and health professional by the company and corporate officials is not uncommon.

The hearing is conducted in the same manner as any other criminal hearing. Normally, each individual and the corporations charged will have separate hearings. Usually, separate legal counsel is required for each of the parties. The burden of proof is on the state to prove each element of the charge *beyond a shadow of a doubt* that the individual committed the offenses alleged and to the degree alleged. All Constitutional rights of a jury trial, cross examination, equal treatment, and self-incrimination are the same as any other criminal trial. If convicted, appeal rights are normally preserved and sentencing would be in accordance with the state's criminal code.

The defenses available in this type of criminal action are varied depending on the situation and the charges. The common law defenses of duress, self defense, defense of others, defense of property, consent, and entrapment are available in addition to the usual defenses to the particular charge. Due to the fact that the injury or fatality occurred on the job, defenses of double jeopardy may be available if previously cited by the OSHA state plan in the same state and preemption of jurisdiction by OSHA. In addition to the factual defenses, other defenses may include, but not be limited to, the application of OSHA or state plan standard, lack of employment relationship,[1] isolated incident defense,[2] or even the lack of employer knowledge defense.[3]

A peripheral area of concern for safety and health professionals following a workplace fatality, major workplace injury, or extensive property damage incident is the potential efficacy losses (i.e., reputation, image, etc.) that can result through widespread media distribution of information. Safety and health professionals should control the information available to the media and attempt to minimize photographs, videotape footage, and other documentation by these outside parties. Failure to control this flow of information and documentation can often result in insurmountable damage at a later time. For example, if the media should acquire the name of the employee who was killed in a work-related accident and announce this information prior to the safety and health professional (or designated person) contacting the family, there is a substantial likelihood that this impersonal method of notification to the family may cause the family members immense pain, the company's reputation could be perceived as being of an uncaring nature, and the relationship between the family and the safety and health professional or company could be permanently harmed.

In summation, safety professionals possess "the right to remain silent…everything you say can and will be used against you in a court of law …" (Miranda Warning). After a disaster situation where employees or others have been severely injured or killed, safety professionals should expect not only a visit from OSHA and other governmental agencies but also from the local law enforcement agency and/or prosecutor. It is important that safety professionals understand the rules when dealing in the criminal law arena versus the civil law arena. Appropriate preparedness within the structure of the emergency and disaster preparedness plan should address issues involving legal counsel and knowledge of legal protections for members of the management team. Remember, you have a right to counsel, and everything you say and do is subject to use against you in a court of law. Be prepared and safeguard your personal rights and the rights of your company.

---

[1] *Gilles & Cotting, Inc*, 1 OSH Cases 1388, **rev'd sub nom.**, *Brennan v. Gilles & Cotting, Inc.*, 504 F.2d 1255 (4th Cir.), **on remand**, 3 OSH Case 2002 (1976).
[2] *Brennan v. Butler Lime and Cement Co.*, 520 F. 2d 1011 (7th Cir. 1975).
[3] *Brennan v. OSHRC*, 511 F.2d 1139 (9th Cir. 1975).

appendix A

OSHA inspection checklist

The following is a recommended checklist for safety professionals in order to prepare for an OSHA inspection:

1. Assemble a team from the management group and identify specific responsibilities in writing for each team member. The team members should be given appropriate training and education and should include, but not be limited to:
   a. an OSHA inspection team coordinator
   b. a document control individual
   c. individuals to accompany the OSHA inspector
   d. a media coordinator
   e. an accident investigation team leader (where applicable)
   f. a notification person
   g. a legal advisor (where applicable)
   h. a law enforcement coordinator (where applicable)
   i. a photographer
   j. an industrial hygienist
2. Decide on and develop a company policy and procedure to provide guidance to the OSHA inspection team.
3. Prepare an OSHA inspection kit, including all equipment necessary to properly document all phases of the inspection. The kit should include equipment such as a camera (with extra film and batteries), a tape player (with extra batteries), a video camera, pads, pens, and other appropriate testing and sampling equipment (i.e., a noise level meter, an air sampling kit, etc.).
4. Prepare basic forms to be used by the inspection team members during and following the inspection.
5. When notified that an OSHA inspector has arrived, assemble the team members along with the inspection kit.
6. Identify the inspector. Check his or her credentials and determine the reason for and type of inspection to be conducted.

7. Confirm the reason for the inspection with the inspector (targeted, routine inspection, accident, or in response to a complaint).
   a. For a random or target inspection
      – Did the inspector check the OSHA 200 Form?
      – Was a warrant required?
   b. For an employee complaint inspection
      – Did inspector have a copy of the complaint? If so, obtain a copy.
      – Do allegations in the complaint describe an OSHA violation?
      – Was a warrant required?
      – Was the inspection protested in writing?
   c. For an accident investigation inspection
      – How was OSHA notified of the accident?
      – Was a warrant required?
      – Was the inspection limited to the accident location?
   d. If a warrant is presented
      – Were the terms of the warrant reviewed by local counsel?
      – Did the inspector follow the terms of the warrant?
      – Was a copy of the warrant acquired?
      – Was the inspection protested in writing?
8. Assess the opening conference.
   a. Who was present?
   b. What was said?
   c. Was the conference taped or otherwise documented?
9. Check records.
   a. What records were requested by the inspector?
   b. Did the document control coordinator number the photocopies of the documents provided to the inspector?
   c. Did the document control coordinator maintain a list of all photocopies provided to the inspector?
10. Review the facility inspection.
    a. What areas of the facility were inspected?
    b. What equipment was inspected?
    c. Which employees were interviewed?
    d. Who was the employee or union representative present during the inspection?
    e. Were all the remarks made by the inspector documented?
    f. Did the inspector take photographs?
    g. Did a team member take similar photographs?[1]

---

[1] "Preparing for an OSHA Inspection," Schneid, T., *Kentucky Manufacturer*, Feb., 1992.

appendix B

Employee workplace rights

Introduction

The Occupational Safety and Health (OSHA) Act of 1970 created the Occupational Safety and Health Administration (OSHA) within the Department of Labor and encouraged employers and employees to reduce workplace hazards and to implement safety and health programs. In so doing, this gave employees many new rights and responsibilities, including the right to do the following:

- Review copies of appropriate standards, rules, regulations, and requirements that the employer should have available at the workplace
- Request information from the employer on safety and health hazards in the workplace, precautions that may be taken, and procedures to be followed if the employee is involved in an accident or is exposed to toxic substances
- Have access to relevant employee exposure and medical records
- Request the OSHA area director to conduct an inspection if they believe hazardous conditions or violations of standards exist in the workplace
- Have an authorized employee representative accompany the OSHA compliance officer during the inspection tour
- Respond to questions from the OSHA compliance officer, particularly if there is no authorized employee representative accompanying the compliance officer on the inspection "walkaround"
- Observe any monitoring or measuring of hazardous materials and see the resulting records, as specified under the Act, and as required by OSHA standards
- Have an authorized representative, or review themselves, the Log and Summary of Occupational Injuries (OSHA No. 200) at a reasonable time and in a reasonable manner
- Object to the abatement period set by OSHA for correcting any violation in the citation issued to the employer by writing to the OSHA area director within 15 working days from the date the employer receives the citation

Submit a written request to the National Institute for Occupational Safety and Health (NIOSH) for information on whether any substance in the workplace has potentially toxic effects in the concentration being used, and have their names withheld from the employer, if so requested

Be notified by the employer if the employer applies for a variance from an OSHA standard, and testify at a variance hearing, and appeal the final decision

Have their names withheld from employer, upon request to OSHA, if a written and signed complaint is filed

Be advised of OSHA actions regarding a complaint and request an informal review of any decision not to inspect or to issue a citation

File a section 11(c) discrimination complaint if punished for exercising the above rights or for refusing to work when faced with an imminent danger of death or serious injury and there is insufficient time for OSHA to inspect; or file a section 405 reprisal complaint (under the Surface Transportation Assistance Act (STAA))

Pursuant to section 18 of the Act, states can develop and operate their own occupational safety and health programs under state plans approved and monitored by Federal OSHA. States that assume responsibility for their own occupational safety and health program must have provisions at least as effective as those of Federal OSHA, including the protection of employee rights. There are currently 25 state plans. Twenty-one states and two territories administer plans covering both private and state and local government employment; and two states cover only the public sector. All the rights and responsibilities described in this appendix are similarly provided by state programs.

Any interested person or groups of persons, including employees, who have a complaint concerning the operation or administration of a state plan may submit a Complaint About State Program Administration (CASPA) to the appropriate OSHA regional administrator. Under CASPA procedures, the OSHA regional administrator investigates these complaints and informs the State and the complainant of these findings. Corrective action is recommended when required.

OSHA standards and workplace hazards

Before OSHA issues, amends, or deletes regulations, the agency publishes them in the *Federal Register* so that interested persons or groups may comment.

The employer has a legal obligation to inform employees of OSHA safety and health standards that apply to their workplace. Upon request, the employer must make available copies of those standards and the OSHA law itself. If more information is needed about workplace hazards than the employer can supply, it can be obtained from the nearest OSHA area office.

Under the Act, employers have a general duty to provide work and a workplace free from recognized hazards. Citations may be issued by OSHA when violations of standards are found and for violations of the general duty clause, even if no OSHA standard applies to the particular hazard. The employer also must display in a prominent place the official OSHA poster that describes rights and responsibilities under OSHA's law.

Right to know

Employers must establish a written, comprehensive hazard communication program that includes provisions for container labeling, material safety data sheets, and an employee training program. The program must include a list of the hazardous chemicals in each work area, the means the employer uses to inform employees of the hazards of non-routine tasks (for example, the cleaning of reactor vessels), hazards associated with chemicals in unlabeled pipes, and the way the employer will inform other employers of the hazards to which their employees may be exposed.

Access to exposure and medical records

Employers must inform employees of the existence, location, and availability of their medical and exposure records when employees first begin employment and at least annually thereafter. Employers also must provide these records to employees or their designated representatives, upon request. Whenever an employer plans to stop doing business and there is no successor employer to receive and maintain these records, the employer must notify employees of their right of access to records at least 3 months before the employer ceases to do business. OSHA standards require the employer to measure exposure to harmful substances, the employee (or representative) has the right to observe the testing and to examine the records of the results. If the exposure levels are above the limit set by the standard, the employer must tell employees what will be done to reduce the exposure.

Cooperative efforts to reduce hazards

OSHA encourages employers and employees to work together to reduce hazards. Employees should discuss safety and health problems with the employer, other workers, and union representatives (if there is a union). Information on OSHA requirements can be obtained from the OSHA area office. If there is a state occupational safety and health program, similar information can be obtained from the state.

OSHA state consultation service

If an employer, with the cooperation of employees, is unable to find acceptable corrections for hazards in the workplace, or if assistance is needed to identify hazards, employees should be sure the employer is aware of the OSHA-sponsored, state-delivered, free consultation service. This service is intended primarily for small employers in high hazard industries. Employers can request a limited or comprehensive consultation visit by a consultant from the appropriate state consultation service.

OSHA inspections

If a hazard is not being corrected, an employee should contact the OSHA area office (or state program office) having jurisdiction. If the employee submits a written complaint and the OSHA area or state office determines that there are reasonable grounds for believing that a violation or danger exists, the office conducts an inspection.

Employee representative

Under section 8(e) of the Act, the workers' representative has a right to accompany an OSHA compliance officer (also referred to as a compliance safety and health officer, CSHO, or inspector) during an inspection. The representative must be chosen by the union (if there is one) or by the employees. Under no circumstances may the employer choose the workers' representative.

If employees are represented by more than one union, each union may choose a representative. Normally, the representative of each union will not accompany the inspector for the entire inspection, but will join the inspection only when it reaches the area where those union members work.

An OSHA inspector may conduct a comprehensive inspection of the entire workplace or a partial inspection limited to certain areas or aspects of the operation.

Helping the compliance officer

Workers have a right to talk privately to the compliance officer on a confidential basis whether or not a workers' representative has been chosen.

Workers are encouraged to point out hazards, describe accidents or illnesses that resulted from those hazards, describe past worker complaints about hazards, and inform the inspector if working conditions are not normal during the inspection.

Observing monitoring

If health hazards are present in the workplace, a special OSHA health inspection may be conducted by an *industrial hygienist*. This OSHA inspector may take samples to measure levels of dust, noise, fumes, or other hazardous materials.

OSHA will inform the employee representative as to whether the employer is in compliance. The inspector also will gather detailed information about the employer's efforts to control health hazards, including results of tests the employer may have conducted.

Reviewing OSHA Form 200

If the employer has more than 10 employees, the employer must maintain records of all work-related injuries and illnesses, and the employees or their representative have the right to review those records. Some industries with very low injury rates (e.g., insurance and real estate offices) are exempt from recordkeeping.

Work-related minor injuries must be recorded if they resulted in restriction of work or motion, loss of consciousness, transfer to another job, termination of employment, or medical treatment (other than first-aid). All recognized work-related illnesses and non-minor injuries also must be recorded.

After an inspection

At the end of the inspection, the OSHA inspector will meet with the employer and the employee representatives in a closing conference to discuss the abatement of hazards that have been found. If it is not practical to hold a joint conference, separate conferences will be held, and OSHA will provide written summaries, on request.

During the closing conference, the employee representative may describe, if not reported already, what hazards exist, what should be done to correct them, and how long it should take. Other facts about the history of health and safety conditions at the workplace may also be provided.

Challenging abatement period

Whether or not the employer accepts OSHA's actions, the employee (or representative) has the right to contest the time OSHA allows for correcting a hazard. This contest must be filed in writing with the OSHA area director within 15 working days after the citation is issued. The contest will be decided by the Occupational Safety and Health Review Commission. The Review Commission is an independent agency and is not part of the Department of Labor.

Variances

Some employers may not be able to comply fully with a new safety and health standard in the time provided due to shortages of personnel, materials, or equipment. In situations like these, employers may apply to OSHA for a temporary variance from the standard. In other cases, employers may be using methods or equipment that differ from those prescribed by OSHA, but which the employer believes are equal to or better than OSHA's requirements, and would qualify for consideration as a permanent variance. Applications for a permanent variance must basically contain the same information as those for temporary variances.

The employer must certify that workers have been informed of the variance application, that a copy has been given to the employee's representative, and that a summary of the application has been posted wherever notices are normally posted in the workplace. Employees also must be informed that they have the right to request a hearing on the application.

Employees, employers, and other interested groups are encouraged to participate in the variance process. Notices of variance application are published in the Federal Register inviting all interested parties to comment on the action.

Confidentiality

OSHA will not tell the employer who requested the inspection unless the complainant indicates that he or she has no objection.

Review if no inspection is made

The OSHA area director evaluates the complaint from the employee or representative and decides whether it is valid. If the area director decides not to inspect the workplace, he or she will send a certified letter to the complainant explaining the decision and the reasons for it. Complainants must be informed that they have the right to request further clarification of the decision from the area director; if still dissatisfied, they can appeal to the OSHA regional administrator for an informal review. Similarly, a decision by an area director not to issue a citation after an inspection is subject to further clarification from the area director and to an informal review by the regional administrator.

Discrimination for using rights

Employees have a right to seek safety and health on the job without fear of punishment. That right is spelled out in section 11(c) of the Act. The law says the employer *shall not* punish or discriminate against employees for exercising such rights as complaining to the employer, union, OSHA, or any other government agency about job safety and health hazards; or for participating in OSHA inspections, conferences, hearings, or other OSHA-related activities.

Although there is nothing in the OSHA law that specifically gives an employee the right to refuse to perform an unsafe or unhealthful job assignment, OSHA's regulations, which have been upheld by the U.S. Supreme Court, provide that an employee may refuse to work when faced with an imminent danger of death or serious injury. The conditions necessary to justify a work refusal are very stringent, however, and a work refusal should be an action taken only as a last resort. If time permits, the unhealthful or unsafe condition should be reported to OSHA or other appropriate regulatory agency.

A state that is administering its own occupational safety and health enforcement program pursuant to section 18 of the Act must have provisions

Appendix B: Employee workplace rights

as effective as those of section 11(c) to protect employees from discharge or discrimination. OSHA, however, retains its section 11(c) authority in all states regardless of the existence of an OSHA-approved state occupational safety and health program.

Workers believing they have been punished for exercising safety and health rights must contact the nearest OSHA office within 30 days of the time they learn of the alleged discrimination. A representative of the employee's choosing can file the 11(c) complaint for the worker. Following a complaint, OSHA will contact the complainant and conduct an indepth interview to determine whether an investigation is necessary.

If evidence supports the conclusion that the employee has been punished for exercising safety and health rights, OSHA will ask the employer to restore that worker's job, earnings, and benefits. If the employer declines to enter into a voluntary settlement, OSHA may take the employer to court. In such cases, an attorney of the Department of Labor will conduct litigation on behalf of the employee to obtain this relief.

Section 405 of the Surface Transportation Assistance Act was enacted on January 6, 1983, and provides protection from reprisal by employers for truckers and certain other employees in the trucking industry involved in activity related to commercial motor vehicle safety and health. Secretary of Labor's Order No. 9-83 (48 *Federal Register* 35736, August 5, 1983) delegated to the Assistant Secretary of OSHA the authority to investigate and to issue findings and preliminary orders under section 405.

Employees who believe they have been discriminated against for exercising their rights under section 405 may file a complaint with OSHA within 180 days of the discrimination. OSHA will then investigate the complaint, and within 60 days after it was filed, issue findings as to whether there is a reason to believe section 405 has been violated.

If OSHA finds that a complaint has merit, the agency also will issue an order requiring, where appropriate, abatement of the violation, reinstatement with back pay and related compensation, payment of compensatory damages, and the payment of the employee's expenses in bringing the complaint. Either the employee or employer may object to the findings. If no objection is filed within 30 days, the finding and order are final. If a timely filed objection is made, however, the objecting party is entitled to a hearing on the objection before an Administrative Law Judge of the Department of Labor.

Within 120 days of the hearing, the Secretary will issue a final order. A party aggrieved by the final order may seek judicial review in a court of appeals within 60 days of the final order. The following activities of truckers and certain employees involved in commercial motor vehicle operation are protected under section 405:

> Filing of safety or health complaints with OSHA or other regulatory agency relating to a violation of a commercial motor vehicle safety rule, regulation, standard, or order

Instituting or causing to be instituted any proceedings relating to a violation of a commercial motor vehicle safety rule, regulation, standard or order

Testifying in any such proceedings relating to the above items.

Refusing to operate a vehicle when such operation constitutes a violation of any federal rules, regulations, standards or orders applicable to commercial motor vehicle safety or health; or because of the employee's reasonable apprehension of serious injury to himself or the public due to the unsafe condition of the equipment

Complaining directly to management, coworkers, or others about job safety or health conditions relating to commercial motor vehicle operation

Complaints under section 405 are filed in the same manner as complaints under 11(c). The filing period for section 405 is 180 days from the alleged discrimination, rather than 30 days as under section 11(c).

In addition, section 211 of the Asbestos Hazard Emergency Response Act provides employee protection from discrimination by school officials in retaliation for complaints about asbestos hazards in primary and secondary schools.

The protection and procedures are similar to those used under section 11(c) of the OSHA Act. Section 211 complaints must be filed within 90 days of the alleged discrimination.

Finally, section 7 of the International Safe Container Act also provides employee protection from discrimination in retaliation for safety or health complaints about intermodal cargo containers designed to be transported interchangeably by sea and land carriers. The protection and procedures are similar to those used under section 11(c) of the OSHA Act. Section 7 complaints must be filed within 60 days of the alleged discrimination.

Employee responsibilities

Although OSHA does not cite employees for violations of their responsibilities, each employee "shall comply with all occupational safety and health standards and all rules, regulations, and orders issued under the Act" that are applicable. Employee responsibilities and rights in states with their own occupational safety and health programs are generally the same as for workers in states covered by federal OSHA. An employee should do the following:

Read the OSHA Poster at the jobsite

Comply with all applicable OSHA standards

Follow all lawful employer safety and health rules and regulations, and wear or use prescribed protective equipment while working

Report hazardous conditions to the supervisor

Report any job-related injury or illness to the employer, and seek treatment promptly

Cooperate with the OSHA compliance officer conducting an inspection if he or she inquires about safety and health conditions in the workplace

Exercise rights under the Act in a responsible manner

Contacting NIOSH

NIOSH can provide free information on the potential dangers of substances in the workplace. In some cases, NIOSH may visit a jobsite to evaluate possible health hazards. The address follows:

National Institute for Occupational Safety and Health
Centers for Disease Control
1600 Clifton Road
Atlanta, Georgia 30333
Telephone: 404-639-3061

NIOSH will keep confidential the name of the person who asked for help if requested to do so.

Other sources of OSHA assistance

Safety and health management guidelines

Effective management of worker safety and health protection is a decisive factor in reducing the extent and severity of work-related injuries and illnesses and their related costs. To assist employers and employees in developing effective safety and health programs, OSHA published recommended *Safety and Health Management Program Guidelines* (*Federal Register* 54(18): 3908-3916, January 26, 1989). These voluntary guidelines apply to all places of employment covered by OSHA. The guidelines identify four general elements that are critical to the development of a successful safety and health management program:

- Management commitment and employee involvement
- Worksite analysis
- Hazard prevention and control
- Safety and health training

The guidelines recommend specific actions, under each of these general elements, to achieve an effective safety and health program. A single free copy of the guidelines can be obtained from the OSHA Publications Office, U.S. Department of Labor, OSHA/OSHA Publications, P.O. Box 37535, Washington, D.C. 20013-7535, by sending a self-addressed mail label with your request.

appendix C

Web sites for disaster preparedness

Information and equipment

Eastern Kentucky University
http://www.eku.edu/fse

Kentucky Safety and Health Network, Inc.
http://www.kshn.org

Commerce Business Daily
http://www.cbd.savvy.com

Government Service Administration
http://www.pueblo.gsa.gov

Federal Money Retriever
http://www.idimagic.com

Federal Government Cites
http://fie.com/www/usgov.htm

World Wide Web Library
http://www.law.indiana.edu/law

Healthfinder
http://www.healthfinder.gov

ASSE
http://www.ASSE.org

ASSE, Puget Sound
http://www.wolfnet.com/mroc/asse.html

ASSEEN San Francisco
http://www.midtown.net/sacasse

Army Industrial Hygiene
http://chppm-ww.apges.army.mil/Armvih/

Asbestos Institute
  http://www.odyssee.net/ai

Ergo Web
  http://ergoweb.com

RSI/UK
  http://www.demon.co.uk/rsi

SafetyLine
  http://sa~e.wt.com.au/safetyline

Arbill
  http://www.arbill.com

Brady
  http://www.safetyonline.net/brady

BuilderOnline
  http://www.builderonline.com

CCOHS
  http://www.ccohs.ca/Resources/hshome.htm

Coppus
  http://www.safetyonline.net/coppus

DuPont
  http://www.dupont.com

Eastman
  http://www.e~~stman.com

Kidde
  http://www.netpath.net/kidde

Lion Apparel
  http://www.lionapparel.com

Marshall
  http://www.marshall.com

MSA
  http://www.msanet.com

Peltor
  http://segwun.muskoka.net/erl/pelter/html

Red Cross
  http://www.redcross.org

Reg Scan
  http://www.regscan.com

Safeware
  http://www.safetyonline.net/safeware

Seaton
http://www.seton.com/directories.html

Strelinger
http://www.strelinger.com

3M
http://www.mmm.com

USC
http://www.usc.eduldept/issm/SH.html

Uvex
http://www.evux.com

National Safety Council
http://www.nsc.org/nsc
http://www.cais.com/nsc

Safety Online
http://www.safetyonline.net

Job Stress Network
http://www.serve.net/cse

MSDS, University of Utah
gopher:/Aitlas.chem.utah.edu:7O/1 1/MSDS

American Health Consultants
http://www.ahcpub.com

World Health Organization
http://www.who.ch

Operation Safe Site
http://www.opsafesite/com

Arthur D. Little
http://www.adlittle.com

Cabot Safety Corp.
http://www.cabotsafety.com

First Aid Direct
http://www.first-aid.com

Fisher Scientific
http://www.fisherl.com

H.L. Bouton Co. Inc.
http://www.safetyonline.netlbouton

Lab Safety Supply
http://www.labsafety.com

MicroClimate Systems
http://www.microclimate.com

CMC Rescue Equipment
http://www.cmcrescue.com/

Able Ergonomics Corp
http://www.ableworks.com

J.J. Keller & Associates
http://www.iikeller.com/keller.html

W.W. Grainger Inc.
http://www.grainger.com

Pro-Am Safety Inc.
http://www~pro-am.com

Vallen Safety Supply
http://www.vallen.com

Ansell Edmont Industrial Inc.
http://www.industry.net/ansell.edmont

Conney Safety Products Co.
http://www.safetyonline.net/conney

Steel Structures Painting Council
http://www.sspc.org

Typing Injuries
http://alumni.caltech.edu/dank/typin-archive.html

Compliance Control Center
http://users.aol.com/comcontrol/comply.html

Pathfinder Associates
http://www.webcom.com/pathfinder/welcome.html

Timber Falling Consultants
http://www.empnet.com/DentD/docs/internet.htm

Safety Directors' Home Page
http://www.unf.edu/iweeks

Penn State University
http://www.ennr.psu.edu/www/dept/arc/server/wikaerob.html

Coastal Video Communications Corp
http://www.safetyonline.net/coastal

Occupational Safety Services Inc.
http://www.k2nesoft.com/ossinc

Enviro-Net MSDS Index
http://www.enviro-net.com/msds/msds.html

Appendix C: Web sites for disaster preparedness

University Kansas School of Allied Health
http://www.kumc.edu/SAH

Seton Online Workplace Safety Information
http://www.seton.com/safety.html

University of Virginia's Video Display Ergonomics
http://www.virginia.edu/enhealth/ERGONOMICS/toc.html

National Environmental Safety Compliance
http://www.albany.net/nesc

American Society of Mechanical Engineers
http://www.asme.org

Canadian Centre for Occ. Health and Safety
http://www.ccohs.ca/Resources/hshome.html

Denison University, Campus Security and Safety
http://www.denison.edu/sec-safe

Duke University Occupational & Environmental Medicine
http://occ-env-med.mc.duke.edu/oem

National Technical Information Service
http://www.fedworld.govntis/ntishome/html

Institute for Research in Construction
http://www.cisti.nrc.ca/irc/irccontents.html

MSU Radiation, Chemical, & Biologcal Safety
http://www.orcbs msu.edu

Rocky Mountain Center for Occupational and Environmental Health
http://rocky.utah.edu

TrainingNet, Trench Safety
http://www.auburn.edu/academic/architecture/bsc/research/trench/index.html

University of Iowa Institute for Rural & Environmental Health
http://info.pmeh.uiowa.edu

University of London Ergonomics & Human Computer Interaction
http://wvw.eroohci.ucl.ac.uk

University of Virginia EPA Chemical Substance Factsheets
http://ecosyst.drdr.virginia.edu/11/1ibrarv/gen/toxics

Government Web Sites Search Engine
http://www.jefflevy.com/gov.htm

National Institutes of Health
http://www.nih.gov

NIOSH
http://www.cdc.gov/niosh/Iiomepag.html

FEMA
http://www.fema.gov/femahndex.html

OSHA
http://www.osha.gov

FedWorld
http://www.fedworld.gov

CDC
http://www.cdc.gov

Federal Agencies
http://www.lib.lsu.edu/gov/fedgov.html

U.S. Department of Labor
http://www.dol.gov/cgi-bin/consolid.pl?media+press

Federal Register
http://www.access.gpo.gov/su docs/aces/aces/40.html

Agency for Toxic Substances and Disease Registry
http://atsdr1.atsdr.cdc.gov:8O8O/atsdrhome.html

U.S. Department of Health and Human Services
http://www.os.dhhs.gov

NIST
http://www.nist.gov/welcome.html

appendix D

Typical responsibilities

Introduction

This appendix identifies emergency-related responsibilities that typically are assigned to the individuals and organizations that are listed below.

Individuals

Chief Executive Officer
Emergency Manager
EOC Manager
Communications Coordinator
Evacuation Coordinator
Mass Care Coordinator
Mass Care Facility Manager
Public Health Manager
Public Information Officer
Resource Manager
Warning Coordinator
County Coroner/Medical Examiner
Agricultural Extension Agent

Departments and agencies

Fire Department
Police Department
Emergency Medical Services
Public Works
Education Department (Superintendent of Education)
Legal Department
Military Department
Mental Health Agencies
Animal Care and Control Agency

Comptroller's Office (or equivalent)
Department of General Services (or equivalent)
Office of Personnel, Job Service
Office of Economic Planning (or equivalent)
Department of Transportation (or equivalent)

Non-governmental organizations

Health and Medical Facilities
Private Utility Companies
EAS Stations
Local Media Organizations
Volunteer Organizations
American Red Cross (Local)
Salvation Army (Local)
Non-Profit Public Service Organizations

Groups and teams

Communications Section Team Members
Needs Group
Supply Group
Distribution Group

Chief Executive Officer
Direction and Control
- Activate EOP (full or partial activation), when appropriate
- Direct tasked organizations to ensure that response personnel report to the appropriate locations (EOC, emergency scene, work center, staging area, etc.) in accordance with the organization's SOP
- Report to the EOC, when notified
- Provide overall direction of emergency response operations, until an emergency scene is established and a field commander assumes this responsibility
- Designate a field commander to direct tactical operations at each emergency scene, as appropriate
- Direct implementation of protective actions for public safety, as appropriate
- Direct EOC staff to relocate to the alternate EOC to continue operations, if necessary
- Terminate response operations and release personnel, when appropriate

Communications
- Require the communications coordinator to report to the EOC when notified of an emergency situation

Appendix D: Typical responsibilities 191

Warning
- Specify who has authority to order the activation of warning systems, to include EAS
- Assign a single organization (normally the EOC) the responsibility for activation of the various warning systems in the jurisdiction; the organization must be able to initiate the warning systems around-the-clock
- Designate public service agencies, personnel, equipment, and facilities that can augment the jurisdiction's warning capabilities

Emergency Public Information
- Serve as primary spokesperson before media, or delegate function to the PIO
- Give final approval to the release of emergency instructions and information, or delegate the function to the PIO
- Designate the location for media briefings (e.g., EOC conference room)
- Approve the implementation of any special provisions for media convergence

Evacuation
- Require the evacuation coordinator to report to the EOC when notified of an emergency situation
- Issue a statement on the jurisdiction's policy on people that do not comply with an evacuation order; the statement addresses the consequences for not evacuating and the services (food, medical, utilities, sanitation, etc.) that will be discontinued or interrupted in the evacuation area
- Issue an evacuation order, when appropriate

Mass Care
- Require the mass care coordinator to report to the EOC when notified of an emergency condition
- Issue an order to open mass care facilities, when appropriate

Health and Medical
- Require the Public Health Officer to report to the EOC when notified of an emergency situation

Emergency Program Manager
Direction and Control
- Ensure that appropriate staff members report to the EOC. Duties may include:
  - Coordinating EOC operations
  - Staffing the Information Processing section
  - Advising/briefing the CEO and other key members of the emergency response organization on the emergency situation

- Recommending to the CEO actions to protect the public from the life threatening consequences associated with the emergency situation

Emergency Public Information
- Advise the CEO on when to disseminate emergency instructions to the public
- Assist the PIO with news releases and rumor control

Evacuation
- Make recommendations to the CEO on the appropriate evacuation option to implement
- Ensure that functional coordinators are clear on the location(s) of Mass Care facilities outside of the risk area that will be used to house evacuees
- Coordinate with and assist the Animal Care and Control Agency staff to identify facilities that may be used to house evacuated animals

Mass Care
- Make recommendations to the CEO on the number and the location(s) of the mass care facilities to be opened
- Coordinate with the PIO to facilitate dissemination of information to the public on the location(s) of the mass care facilities that will be opened
- Coordinate with the Mass Care Coordinator to activate the jurisdictions' mass care facilities

Resource Management
- Assist the Resource Manager as necessary during response operations

EOC Manager (usually staffed by the Emergency Manager)

Direction and Control
- Immediately, notify the CEO of significant emergency situations that could affect the jurisdiction
- When directed by the CEO, or when circumstances dictate, notify all tasked organizations, inform them of the situation, and direct them to take the action appropriate for the situation (report to the EOC or the scene of the emergency, stand by, etc.) in accordance with their organization's SOP
- Activate the EOC when directed to do so by the CEO or when the situation warrants such action
- Manage EOC resources and direct EOC operations. Duties may include ensuring that the following activities/actions are done:
  - Information processing (collecting, evaluating, displaying, and disseminating information about the emergency situation to help

support the jurisdiction's response operations). Typical tasks may include:
- Maintaining a significant events log
- Message handling
- Aggregating damage information from all available sources
- Identifying resource needs
- Preparing summaries on the status of damage
- Preparing briefings for senior management officials
- Displaying appropriate information in the EOC
- Preparing and submitting necessary reports when required (reports on the situation, critical resource status, etc.)

– Coordinate logistical support for response personnel and disaster victims
– When directed by the CEO, or when conditions warrant such action, relocate staff to the alternate EOC to continue response operations
– When directed by the CEO, terminate operations and closing the EOC

Communications
- Activate the communications section in the EOC
- Implement emergency communications procedures
- Ensure that the communications section of the EOC has the capability to sustain operations around the clock

Warning
- Activate the warning section in the EOC
- Ensure that emergency warning systems are activated when directed to do so
- Issue cancellation of warning notices or otherwise ensure that emergency responders and the public are aware of the fact that the emergency situation is terminated

Communications Coordinator
Direction and Control
- Serve as a member of the EOC team
- Ensure that the emergency communications section in the EOC is equipped with the appropriate communication gear

Communications
- When notified of an emergency, report to the EOC, if appropriate
- Manage the emergency communications section in the EOC and supervise the personnel (radio, telephone and teletype operators, repair crews, runners, etc.) assigned to it
- Support media center communications operations as necessary

Evacuation Coordinator

Direction and Control
- When notified of an emergency situation, report to the EOC, if appropriate
- Coordinate the implementation of evacuation actions for humans with the appropriate tasked organizations

Evacuation
- Review known information about the emergency situation and make recommendations to the Emergency Program Manager on the appropriate evacuation options to implement
- Identify assembly areas for picking up people who do not have their own transportation
- Identify evacuation routes
- Select primary routes from the risk area to designated mass care facilities
- Determine the loading potential of each primary route
- Examine access to primary routes from each part of the risk area
- Prepare the evacuation movement control plan
- Assist, as appropriate, the Animal Care and Control Agency's efforts to evacuate animals at risk during catastrophic emergency situations

Mass Care Coordinator

Direction and Control
- When notified of an emergency situation, report to the EOC, if appropriate
- Coordinate the implementation of mass care actions for humans with the appropriate tasked organizations

Evacuation
- Activate staff and open mass care facilities outside of the evacuation area when directed to do so by appropriate authority

Mass Care
- Assess the situation and make recommendations to the Emergency Manager on the number and locations of mass care facilities to be opened
- Review the listing of available mass care facilities
- Notify persons and organizations identified in the mass care resource list about the possible need for services and facilities
- Select mass care facilities for activation in accordance with:
  - Hazard/vulnerability analysis considerations
  - Locations in relation to evacuation routes
  - Services available in facilities
- When directed, coordinate the necessary actions to ensure that mass care facilities are opened and staffed, as necessary

- Notify mass care facility managers to do one of the following, when appropriate:
  - Stand by for further instruction on the specific actions to take and the estimated timing for opening mass care facilities
  - Take the necessary action to open the facility they are responsible for managing
- Ensure that each mass care facility receives its supplies
- Coordinate with EOC staff to ensure that communications are established, routes to the mass care facilities are clearly marked, and appropriate traffic control systems are established
- Ensure that each mass care facility has a highly visible identity marker and a sign that identifies its location
- Provide each mass care facility manager with a listing of the location of the animal shelters that have been opened to house and care for companion animals
- Assist, as appropriate, the Animal Care and Control Agency's efforts to feed, shelter, and provide medical treatment for animals during catastrophic emergencies
- Upon termination of the emergency, submit a mass care expenditure statement to appropriate authorities for reimbursement

Mass Care Facility Manager

Mass Care
- When notified, stand by for further instructions or report to the assigned mass care facility, as appropriate
- Contact team members and instruct them to take whatever actions may be appropriate
- Staff and operate the mass care facility; upon arrival at the facility, take the necessary actions to open it, receive evacuees, and provide for their health and welfare
- Contact the EOC when the facility is ready to open
- Open and keep the facility operating as long as necessary
- Implement registration and inquiry procedures for all evacuees who enter the facility
- Ensure that individual and family support services are provided at the mass care facility
- If companion animals are not permitted in the facility, help their owners to place them in a shelter that has been opened to house and care for animals
- Each day, report the following to the EOC:
  - Number of people staying in the facility
  - Status of supplies
  - Condition of the facility and any problem areas
  - Requests for specific types of support, as necessary
- Maintain records of expended supplies

- Arrange for the return of evacuees to their homes or for transportation to long-term temporary shelter, if necessary
- When appropriate, terminate operations and close the facility
- Clean the facility and return it to its original condition
- Submit a mass care facility status report to the Mass Care Coordinator, identifying the equipment and supplies that are needed to restock the facility and any other problems that will need to be resolved before the facility is used again

Public Health Coordinator
Direction and Control
- When notified of an emergency situation, send a representative to the EOC, if appropriate
- Coordinate the medical treatment activities of all response organizations involved in providing medical assistance to disaster victims
- Coordinate the necessary mortuary services, to include operations of temporary morgues, and identification of victims
- Collect information and report damage and the status of health and medical facilities and equipment to the EOC

Health and Medical
- Ensure that the emergency medical teams that responded to the disaster site establish a medical on-scene command post and that a single individual is in charge of all medical operations
- Coordinate the use of all public health services in the jurisdiction during emergency conditions
- Coordinate health-related activities among other local public and private response agencies or groups
- Coordinate with the neighboring areas and state public health officers on matters that require assistance from other jurisdictions
- Provide for the monitoring and evaluation of environmental health risks or hazards as required and ensure that the appropriate actions are taken to protect public safety
- Inspect for purity and usability of foodstuffs, water, drugs, and other consumables that were exposed to the hazard
- Detect and inspect sources of contamination that are dangerous to the general public's physical and mental health
- Inspect damaged buildings for health hazards
- Provide epidemiological surveillance, case investigating, and follow-up
- Provide laboratory services for the identification required to support emergency health and emergency medical services
- Implement action to prevent or control vectors such as flies, mosquitoes, and rodents, and work with animal care and control agencies to prevent the spread of disease by animals

- Coordinate operations for general or mass emergency immunizations or quarantine procedures
- Establish preventive health services, including the control of communicable diseases
- Coordinate with the water, public works, or sanitation departments, as appropriate, to ensure the availability of potable water and an effective sewage system, sanitary garbage disposal, and the removal of dead animals
- Monitor food handling and mass feeding sanitation service in emergency facilities, including providing increased attention to sanitation in commercial feeding and facilities that are used to feed disaster victims
- Coordinate with the Animal Care and Control Agency to dispose of dead domestic animals and contaminated carcasses
- Provide advice to the public on general sanitation matters. Whenever feasible, all information should be provided to the public and the media through the PIO

County Coroner/Medical Examiner
Health and Medical
- Coordinate local resources utilized for the collection, identification, and disposition of deceased persons and human tissue
- Select an adequate number of qualified personnel to start temporary morgue sites
- Establish collection points to facilitate recovery operations
- Coordinate with search and rescue teams
- Determine cause of death
- Designate an adequate number of persons to perform the duties of Deputy Coroners
- Protect the property and personal effects of the deceased
- Notify the next of kin of the deceased
- Establish and maintain a comprehensive recordkeeping system for continuous updating and recording of fatality numbers
- Submit requests for mutual aid assistance, if required
- Provide information through the PIO to the news media on the number of deaths, morgue operations, etc., as appropriate
- Coordinate the services of:
  - Funeral directors, ambulances, and morticians
  - Other pathologists
  - The American Red Cross for location and notification of relatives
  - Dentists and x-ray technicians for purposes of identification
  - Law enforcement agencies for security, property protection, and evidence collection
  - Mutual aid provision to other counties upon request

Emergency Medical Services
Health and Medical
- Respond to the disaster scene with emergency medical units (rescue/ambulance)
- Provide personnel and equipment to administer emergency medical assistance at the disaster scene
- Coordinate with hospitals and other public health services organizations to ensure that all medical operations are thoroughly integrated
- Assist in the triage of the injured, as appropriate
- Implement a medical incident management system, such as the Incident Command System, within the overall framework of the jurisdiction's emergency management system
- Coordinate with local and regional hospitals to ensure that casualties are transported to the appropriate hospital
- Provide appropriate emergency first aid/medical supplies for disaster use
- Maintain updated resource inventories of emergency medical supplies and equipment
- Maintain a casualty/patient tracking system
- Establish and maintain field communications and coordination with other responding emergency teams (medical, fire, police, public works, etc.), and radio or telephone communication with hospitals, as appropriate
- Maintain liaison with the American Red Cross and volunteer service agencies within the jurisdiction
- Coordinate with business and industry emergency medical units
- Coordinate the procurement, screening, and allocation of critical public and private resources that are required to support disaster-related health and medical care operations
- If appropriate, provide information through the PIO to the news media on the number of injuries, deaths, etc.

Mental Health Agencies
Evacuation
- Coordinate the evacuation of patients from damaged or threatened mental health facilities
- Prepare for and coordinate the reception of patients evacuated from other such facilities

Health and Medical
- Ensure that professional psychological support is available for victims and involved personnel (on an as-needed basis) during all phases of the disaster

Appendix D: Typical responsibilities 199

- At inpatient facilities:
  - Care for patients who reside in mental health facilities
  - Implement the mental health facility disaster plan
  - Protect and provide security for those people committed to inpatient mental health facilities

Health and Medical Facilities

Evacuation
- Reduce the patient population in hospitals, nursing homes, and other health care facilities, if evacuation becomes necessary
- Provide transport and medical care for the patients that are being evacuated
- Provide continued medical care for patients who cannot be moved when hospitals, nursing homes, and other health care facilities are evacuated

Health and Medical
- Implement the hospital's disaster plan
- Establish and maintain field and inter-hospital medical communications
- Provide medical guidance, as necessary, to EMS units, field collection and/or treatment locations, etc.
- Coordinate with medical response personnel at the disaster scene to ensure that casualties are transported to the appropriate medical facility
- Distribute patients to and among hospitals based on capability to treat and bed capacity, including transfers out of the area and/or rerouting to alternative facilities
- Make available upon request qualified medical personnel, supplies, and equipment located in the jurisdiction
- Coordinate with other hospitals involved in caring for the injured
- Maintain liaison with the coordinators of the emergency services such as fire and rescue departments, law enforcement, public works, emergency management agency, etc.
- If appropriate, provide information through the PIO to the news media on the number of injuries, deaths, etc.
- Assist in the reunification of the injured with their families

Public Information Officer

Direction and Control
- When notified, report to the EOC or incident scene, as appropriate
- Handle inquiries and inform the public about disaster damage, restricted areas, actions to protect and care for companion animals, farm animals, and wildlife, and available emergency assistance

Emergency Public Information
- Manage all aspects of EPI on behalf of the CEO
- Assume EPI functions that are delegated by the CEO
- Ensure the timely preparation of EPI materials and their dissemination
- Ensure that the public is able to obtain additional information and provide feedback (e.g., with a hotline for public inquiries)
- Ensure the gathering of necessary information and timely preparation of news releases
- Brief public affairs officers who go to the scene of the emergency
- Schedule news conferences, interviews, and other media access (subject to any special media convergence provisions)
- Supervise the media center
- Assign print and broadcast monitors to review all media reports for accuracy
- Coordinate rumor control activity
- Maintain a chronological record of disaster events

Evacuation
- Disseminate the following types of instructional materials and information to evacuees:
  - Identification of the specific area(s) to be evacuated
  - A list of items that evacuees should take with them (such as food, water, medicines, portable radio, fresh batteries, clothing, sleeping bags)
  - Departure times
  - Pick-up points for people without transportation
  - Evacuation routes
  - Locations of mass care facilities outside of the evacuation area
- Keep evacuees and the general public informed on evacuation activities and the specific actions that they should take
- Disseminate information on appropriate actions to protect and care for companion and farm animals that are to be evacuated or left behind

Mass Care
- Make public announcements about the availability of mass care facilities and animal shelters and their locations

Resource Manager

Direction and Control
- When notified of an emergency, report to the EOC, if appropriate
- Coordinate implementation of resource management activities with the appropriate tasked organizations

Appendix D: *Typical responsibilities*   201

Resource Management
- Direct and supervise the activities of the Needs, Supply, and Distribution Groups
- Coordinate with the EOC Manager and key organizations' representatives in the EOC regarding needs and the priorities for meeting them
- During the emergency, monitor potential resource shortages in the jurisdiction and advise the Emergency Manager or CEO on the need for action
- Identify facilities or sites that may be used to store necessary resources and donations
- Determine the need for, and direct activation of, facilities that are necessary for the coordinated reception, storage, and physical distribution of resources
- Make arrangements for work space and other support needs for Resource Management staff

Needs Group

Resource Management
- Receive requests and report on the function's success in meeting needs; the Needs Group includes the Needs Analyst and Needs Liaisons

NEEDS ANALYST
- When notified of an emergency, report to the EOC, or other location specified by the Resource Manager (RM)
- Tabulate the needs assessment and specific requests
- Prioritize needs for the Supply Group, with the concurrence of the RM
- Provide regular reports to the RM on the status of requests (e.g., pending, en route, met)

NEEDS LIAISONS
- When notified of an emergency, report to the EOC, or other location specified by the RM
- Receive specific requests, eliciting essential information from requesting parties

Supply Group

Resource Management
- Locate and secure resources; this Group is headed by the Supply Coordinator and includes (as necessary) donations, procurement, and personnel teams as well as the Financial Officer and Legal Advisor

SUPPLY COORDINATOR
- When notified of an emergency, report to the EOC, or other location specified by the Resource Manager (RM)
- Determine appropriate means for satisfying requests (with the concurrence of the RM)
- Handle unsolicited bids
- Keep the Needs Group informed of action taken on requests
- Keep the Distribution Group informed of the expected movement of resources, along with the priority designation for the resources
- Request transportation from the Distribution Group (with the concurrence of the RM)

DONATIONS TEAM (headed by the Donations Coordinator)
- When notified of an emergency, report to the EOC, or other location specified by the RM
- Receive offers of donated goods and services
- Match offers to needs (whether those of its own separate needs assessment or those of the larger jurisdictional needs assessment)
- Seek through the PIO to ensure that offers are not inappropriate to needs
- Make special requests as directed by the Supply Coordinator
- Ensure that the RM is apprised of the needs and unmet needs lists and that physical distribution efforts (in those jurisdictions that treat donations logistics separately) are coordinated with the Distribution Group

PROCUREMENT TEAM
- Undertake ad hoc procurement as directed by the Supply Coordinator or use database and/or resource listings to fill requests through prearranged supply channels
- When notified of an emergency, report to the EOC or other location specified by the RM
- When warning is available and as directed by the Supply Coordinator, notify private industry parties to Memorandums of Agreement of the jurisdiction's intent to activate the agreement, confirm the availability of resources specified by the agreement, and reserve a supply
- Locate necessary resources using database and/or resource listings for the jurisdiction and participating suppliers
- As directed by the Supply Coordinator, seek to procure resources that are not available through prearranged channels
- In all cases, contact suppliers, settle terms for transportation, and provide information necessary to pass checkpoints
- Inform the Supply Coordinator when the jurisdiction must provide transportation to make use of the resource

PERSONNEL TEAM
- When notified of an emergency, report to the EOC, or other location specified by the RM
- As directed by the Supply Coordinator, recruit and hire personnel to meet emergency staffing needs

FINANCIAL OFFICER
- When notified of an emergency, report to the EOC or other location specified by the RM
- Oversee the financial aspects of meeting resource requests: record-keeping, budgeting for procurement and transportation, facilitating cash donations to the jurisdiction (if necessary and as permitted by law)

LEGAL ADVISOR
- When notified of an emergency, report to the EOC or other location as specified by the RM
- Advise the Supply Coordinator and Procurement Team on contracts and questions of administrative law

Warning Coordinator
(When practical, this individual should be permanently assigned to the EOC)

Direction and Control
- Develop and maintain a phone and/or frequency list for notifying emergency response personnel, neighboring jurisdictions, and the state of an emergency situation
- Develop and maintain a phone list or other means for warning special locations, such as schools, hospitals, nursing homes, major industrial sites, institutions, and places of public assembly
- Identify public and private service agencies, personnel, equipment, and facilities that could be called upon to augment the jurisdiction's warning capabilities

Warning
- When notified of an emergency, report to the EOC
- Implement call-down rosters to alert emergency responders or provide situation updates
- Activate public warning systems, to include the EAS
- Implement contingency plans to provide warnings if the established warning system fails to work
- Coordinate warning frequencies and procedures with EOCs at higher levels of government and with adjacent jurisdictions
- Work with the PIO to ensure that pertinent warning information is provided to the print media for distribution to the public

Agricultural Extension Agent
Mass Care
- Develop and maintain a list of local food warehouses and other sources of bulk food stocks

Fire Department
Direction and Control
- When notified of an emergency situation, send response teams/personnel, equipment, and vehicles to the emergency site, staging areas, or other locations, as appropriate
- Send a senior representative to the EOC, when the EOC has been activated
- Notify the EOC of the situation if the original source of notification did not come from the EOC
- Perform field commander duties at the emergency scene, if appropriate
- Manage fire and rescue resources, direct fire operations, rescue injured people during emergency operations, and determine the need for evacuation of the immediate area in and around the emergency scene
- Assist in the evacuation of people at risk in the immediate area in and around the emergency scene
- Alert all emergency response organizations of the dangers associated with technological hazards and fire during emergency operations

Police Department
Direction and Control
- When notified of an emergency situation, send response teams/personnel, equipment, and vehicles to the emergency scene or other locations, as appropriate
- Notify the EOC of the situation if the original source of notification did not come from the EOC
- Send a senior representative to the EOC, when the EOC has been activated
- Perform field commander duties at the emergency scene, if appropriate
- Manage law enforcement resources and direct law enforcement operations. Duties may include:
  - Directing and controlling traffic during emergency operations
  - Assisting in the evacuation of people at risk in and around the emergency scene
  - Controlling access to the scene of the emergency or the area that has been evacuated

- Providing security in the area affected by the emergency to protect public and private property
- Conducting damage assessment activity (through use of aircraft, helicopter, or other police vehicles as appropriate)

Evacuation
- Provide traffic control during evacuation operations; operation considerations include:
  - Route assignment departure scheduling
  - Road capacity expansion
  - Entry control for outbound routes
  - Perimeter control on inbound routes
  - Traffic flow, including dealing with breakdowns
  - Establishment of rest areas
  - Securing, protecting, and housing those prisoners that must be evacuated
  - Protection of property in evacuated area
  - Limiting access to the evacuated area

Mass Care
- Provide security and law enforcement at mass care facilities
- Provide traffic control during evacuee movement to mass care facilities
- Maintain order in mass care facilities
- If necessary, provide an alternative communications link between the mass care facility and the EOC through a mobile radio unit in police vehicles

Health and Medical
- Maintain emergency health and environmental health services at juvenile and adult correctional facilities
- Assist the Coroner in the identification of fatalities

Resource Management
- Provide escort and security as appropriate for the delivery, storage, and distribution of resources

Public Works

Direction and Control
- When notified of an emergency situation, send response teams/personnel, equipment, and vehicles to the emergency scene, staging area, or other locations, as appropriate
- Notify the EOC of the situation if the original notification did not come from the EOC
- Send a senior representative to the EOC, when the EOC has been activated

- Perform field commander duties at the emergency scene, if appropriate
- Manage public works resources and direct public works operations; duties may include:
  - Performing debris-removal operations
  - Assisting in urban search and rescue efforts
  - Conducting damage assessment activities (through the use of vehicles, remote video equipment, etc., as appropriate)
  - Providing emergency generators, fuel, lighting, and sanitation to support emergency responders at the emergency scene and at the EOC
  - Coordinating with utility companies to restore power to disaster victims

Evacuation
- Verify the structural safety of routes (roads, bridges, railways, waterways, airstrips, etc.) that will be used to evacuate people

Mass Care
- Ensure that power, water supply, and sanitary services at mass care facilities are maintained during emergency conditions
- Determine the structural safety of mass care facilities as soon as practical after an earthquake has occurred

Education Department (Superintendent of Education)

Direction and Control
- When notified of an emergency, send a representative to the EOC, if appropriate
- Protect students in school when an emergency situation occurs
- Evacuate students, if appropriate
- When directed by appropriate authority, close school facilities and release students
- When directed by appropriate authority, make schools available for use as mass care facilities
- Conduct damage assessment of school facilities

Emergency Public Information
- Disseminate emergency information to school populations as appropriate

Evacuation
- Evacuate students from school buildings when the situation warrants or when directed to do so by the appropriate authority
- Close school facilities and release students from school when directed to do so by the appropriate authority

Mass Care
- If appropriate, provide personnel to manage and staff mass care facilities

Legal Department
Direction and Control
- When notified of an emergency, report to the EOC, if appropriate

Military Department
Direction and Control
- Provide personnel and equipment to support direction and control actions at the scene and/or the EOC (at the direction of the Governor)

Communications
- Provide communications support, to include personnel and equipment (as directed by the Governor)

Mass Care
- Inform the Mass Care Coordinator of mass care facilities that are available on military installations
- Coordinate the use of mass care facilities on military installations
- Provide logistical support for mass care operations

Health and Medical
- Provide personnel and equipment to support medical operations during disaster situations (at the direction of the Governor)

Animal Care and Control Agency
Direction and Control
- When notified of an emergency, send a representative to the EOC, if appropriate
- Manage public and private sector efforts to meet the animal service needs that arise, including:
  - Rescue and capture of animals that have escaped confinement as well as displaced wildlife
  - Evacuation
  - Sheltering
  - Care of injured, sick, and stray animals
  - Disposal of dead animals
- Activate emergency response teams (evacuation, shelter, medical treatment, search and rescue, etc.) as necessary
- Prepare a resource list that identifies the agencies and organizations that are responsible for providing the supplies (medical, food, and

other necessary items) needed to treat and care for injured and sick animals during catastrophic emergencies
- Coordinate response activities with the appropriate representative in the EOC (EOC Manager, Evacuation Coordinator, Mass Care Coordinator, American Red Cross, Public Information Officer, Public Health Services, Resource Manager, etc.)

Evacuation
- Based on information from the Evacuation Coordinator on the high hazard areas in the jurisdiction, make an initial estimate of the numbers and types of animals that may need to be evacuated
- Coordinate with the Evacuation Coordinator to arrange travel routes and schedule the timing for evacuation of farm animals, animals in kennels, veterinary hospitals, zoos, pet stores, animal shelters, university laboratories, etc., and wildlife (as appropriate) from the risk area
- As appropriate, mobilize transportation vehicles (stock trailers, trucks equipped with animal cages, etc.) that may be used to evacuate the animals
- Implement evacuation by sending evacuation teams to load and transport the animals
- As appropriate, dispatch search and rescue teams to look for animals left behind by their owners, stray animals, and others needing transport to a safe location

Mass Care
- Assess the situation and make a decision on the number and location of shelters that will be used to house animals. Typical pet facilities include the jurisdiction's animal shelter(s), veterinary hospitals, boarding kennels, pet stores, greyhound farms, and fairgrounds. Facilities for agricultural animals could include sale barns, boarding stables, race tracks, horse farms, poultry barns, dairy farms, and fairgrounds/rodeo grounds
- Coordinate the actions that are necessary to obtain sufficient personnel to staff animal shelters, as appropriate
- Ensure that each animal shelter has a highly visible identity marker and a sign that identifies its location
- Coordinate with the PIO to facilitate dissemination of information to the public on the location of the companion-animal shelters that will be opened
- Inform the Mass Care Coordinator of the location(s) of the shelters that have been opened
- If appropriate, coordinate with the Mass Care Coordinator to place personnel in public shelters to act as a referral source for disaster operations

- Open shelters and provide food, water, and medical care, as necessary, for the animals in the shelter
- Keep shelters open as long as necessary
- Ensure that each shelter receives the necessary supplies to sustain itself
- When appropriate, terminate shelter operations and close the facility

Health and Medical
- Coordinate with veterinary and animal hospital organizations to arrange for their support and services to help meet the needs of owners of companion animals, farm animals, and wildlife
- Coordinate the rescue of injured and/or endangered wildlife and farm animals with the Fish and Game department, State/National Wildlife officials/organizations, Department of Agriculture, county cooperative extension offices, veterinarians, etc.
- Coordinate with the Public Health Services prior to disposing of dead domestic animals and contaminated carcasses

Comptroller's Office (or equivalent)

Direction and Control
- When notified of an emergency, send a representative to the EOC, if appropriate
- Provide the Resource Manager and the CEO with summary briefings on the status of financial transactions
- Maintain records of all financial transactions during response operations
- Handle all procurement requests initiated by response organizations
- Establish a procedure for the jurisdiction to accept "cash donations," where statute permits such action
- Become familiar with the protocol and procedures required by the Stafford Act that are applicable to reimbursing the jurisdiction for eligible expenses associated with "Presidentially Declared" Disasters
- Upon termination of the response effort, prepare the appropriate reports that address costs incurred by the jurisdiction during the emergency situation

Resource Management
- Provide knowledgeable staff to serve as Financial Officer and associated support

Department of General Services (or equivalent)

Resource Management
- Provide knowledgeable staff to serve on the Supply Group, Distribution Group, and in other capacities as appropriate

Office of Personnel, Job Service

Resource Management
- Provide knowledgeable staff for the Personnel Team to obtain human resources

Office of Economic Planning (or equivalent)

Resource Management
- Provide knowledgeable staff to serve on the Needs Group

Department of Transportation (or equivalent)

Resource Management
- Provide knowledgeable staff to serve on the Distribution Group
- Assist in procuring and providing transportation

Private Utility Companies

Direction and Control
- When notified of an emergency, send a representative to the EOC, if appropriate

EAS Stations

Emergency Public Information
- Store "canned" EPI messages (other than warnings) and disseminate this information at the PIO's request
- Disseminate information when requested to do so by the CEO or his or her designee

Local Media Organizations

Emergency Public Information
- Store/maintain advance emergency packets for release at the PIO's request
- Verify with the PIO field reports of the emergency's development
- Cooperate in public education efforts

Volunteer Organizations

Direction and Control
- When notified of an emergency, send a representative to the EOC, if appropriate

Emergency Public Information
- Provide support to public inquiry phone lines, as requested by the PIO
- Provide support in disseminating printed EPI material, as requested by the PIO

Appendix D: Typical responsibilities 211

American Red Cross (local)
Mass Care
- If appropriate, provide personnel to manage and staff mass care facilities

Health and Medical
- Provide food for emergency medical workers and patients, if requested
- Maintain a medical evacuee tracking system
- Assist in the notification of the next of kin of the injured and deceased
- Assist with the reunification of the injured with their families
- Provide blood, blood substitutes, and blood byproducts, and/or implement reciprocal agreements for the replacement of blood items
- Provide first aid and other related medical support at temporary treatment centers, as requested, and within capability
- Provide supplementary medical nursing aid, and other health services upon request, and within capability
- Provide assistance for the special needs of the handicapped, elderly, orphaned children, and those children separated from their parents

Salvation Army (local)
Mass Care
- If appropriate, provide personnel to manage and staff mass care facilities

Non-profit Public Service Organizations
Mass Care
- If appropriate, provide personnel to manage and staff mass care facilities

Communications Section Team Members
Communications
- When notified of an emergency, report to the EOC and staff the communications section and operate assigned communications equipment
- Follow established procedures and radio protocol for voice transmissions and message handling
- Screen and log information when appropriate, and route incoming calls to the appropriate section in the EOC

Distribution Group
Resource Management
- Ensure the delivery of resources by overseeing routing, transportation, collection, sorting/aggregating, storage, and inventory; this group is headed by the Distribution Coordinator

- When notified of an emergency, report to the EOC or other location specified by the RM
- Transport resources, as requested
- Control the movement of resources
- Perform materials-handling work

DISTRIBUTION COORDINATOR
- Oversee the transportation and physical distribution of resources
- Ensure that facilities are activated and directed by the RM
- Monitor the location, passage, and inventory of resources

appendix E

Potential sources of disaster preparedness and management assistance through local colleges and universities

Local colleges and universities often possess safety, environmental, and loss prevention programs with disaster management and preparedness as part of their curriculum. These centers of higher learning are often a great resource for information and assistance in the development of your programs. Additionally, to assist with the manpower needs in the development of the program, many colleges and universities offer intern and cooperative education programs in which student assistance can be derived.

Auburn University, Leo A. "Tony" Smith, Professor
Industrial Engineering Department
College of Engineering
Auburn University
207 Danston Hall
Auburn, AL 36849-5346
205-844-1415
  BS/MIE/MS/PCD, Industrial Engineering, Concentration in Safety & Ergonomics

The University of Alabama — Tuscaloosa, Dr. Paul S. Ray,
  Assistant Professor
Industrial Engineering Department
The University of Alabama — Tuscaloosa
PO Box 870288
Tuscaloosa, AL 35487-0288
205-346-1603
  BS/MS, Industrial Engineering

University of Alabama — Birmingham, Joan Gennin,
  Program Administrator
Department of Environmental Health Sciences
School of Public Health
University of Alabama — Birmingham
Titmell Hall
Birmingham, AL 35294-0008
205-934-8488
  MPH/PhD/DrPH, Environmental Health Sciences
  MSPH/PhD, Industrial Hygiene
  MPH/DrPH, Occupational Health and Safety
  MSPH/PhD, Environmental Toxicology

Jacksonville State University, J. Fred Williams, Program Director
Department of Technology
Jacksonville State University
Room 217 Self Hall
700 Pelham Road North
Jacksonville, AL 36265
205-782-5080
  Occupational Safety and Health Technology

University of North Alabama, Dr. Robert Gaunder, Professor
Chemistry/Industrial Hygiene
University of North Alabama
UNA Box 5049
Florence, AL 35632
205-760-4474
  BS, Industrial Hygiene

GateWay Community College, Ginger Jackson, Program Director
Industrial Technology Division
Gateway Community College
108 N 40th Street
Phoenix, AZ 85034
602-392-5000
  AA, Occupational Safety and Health

Southern Arkansas University, James A. Collier, Program Head
School of Science and Technology
Southern Arkansas University
100 East University
Magnolia, AR 71753
501-235-4284
  BS, Industrial Technology

Appendix E: Assistance through local colleges and universities 215

University of California — Berkley, Jeanne Bronk, Coordinator
Environmental Health Science Program
School of Public Health
University of California — Berkley
Berkley, CA 94720
510-643-5160
　MS/MPH/PhD, Environmental Health

California State University — Fresno, Dr. Sanford Brown, Advisor
Environmental Health Science Program
California State University — Fresno
2345 E. San Ramon
Fresno, CA 93740-0030
209-278-4747
　BS, Environmental Health

California State University — Fresno, Dr. Michael Waite, Advisor
Occupational Safety and Health Program
California State University — Fresno
2345 E San Ramon
Fresno, CA 93740-0030
209-278-5093
　BS, Occupational Safety and Health

University of Southern California, William J. Petak, Professor
Institute of Safety and Systems Management Building
University of Southern California
University Park
Los Angeles, CA 90089-0021
213-740-2411
　BS/MS, Safety and Health

California State University — Los Angeles, Dr. Carlton Blanton,
　Professor
Health and Science Department
California State University — Los Angeles
5151 State University Drive
Los Angeles, CA 90032
213-343-4740
　BS, Health Science
　BS, Occupational Safety and Health
　MA, Occupational Safety and Health
　Certificate, Occupational Safety and Health
　Certificate, Environmental Health Certificate, Alcohol and Drug
　　Problems

California State University — Northridge, Brian Malec, Chair
Health Science Department
California State University — Northridge
18111 Nordhoft Street
Northridge, CA 91330
818-885-3100
   BS/MS, Environmental Health

Merritt College, Larry Gurley, Assistant Dean
Technical Division
Merritt College
12500 Campus Drive
Oakland, CA 94619
510-436-2409
   AS, Occupational Safety and Health

National University, Ernest Wendi, Program Chair
Management and Technology
Department of Computers and Technology
Suite 205
National University
4141 Camino Del Rio South
San Diego, CA 92108
619-563-7124
   BS/MS, Occupational Health and Safety

Colorado State University, Kenneth Blehm, Coordinator
Department of Environmental Health
College of Veterinary Medicine and Biomed
Colorado State University
Fort Collins, CO 80523
303-491-7038
   BS/MS/PhD, Environmental Health

Red Rocks Community College, Anne-Mario Edwards, Department Coordinator
Department of Occupational Safety Technology
Red Rocks Community College
Campus Bon 41
13300 W 6th Avenue
Lakewood, CO 80401-5398
303-914-6338
   MS, Occupational Safety Technology Certificate, Occupational Safety Technology

Trinidad State Junior College, Charles McGlothlin, Associate Professor
Occupational Safety Department

Appendix E: Assistance through local colleges and universities 217

Trinidad State Junior College
600 Prospect Street
Trinidad, CO 81082
719-846-5502
   AAS, Occupational Safety and Health Certificate, Occupational
   Safety and Health

Central Connecticut State University, Andrew Baron, Assistant Dean
Occupational Safety Health Department
School of Technology
Central Connecticut State University
1615 Stanley Street
New Britain, CT 06050
203-827-7997
   BS, Occupational Safety and Health
   BS, Public Safety

University of New Haven, Dr. Garher, Director
Department of Occupational Safety and Health
University of New Haven
300 Orange Avenue
West Haven, CT 06516
203-932-7175
   AS/BS, Occupational Safety and Health Administration
   AS/BS, Occupational Safety and Health Technology
   MS, Occupational Safety and Health Management
   MS, Industrial Hygiene

Florida International University, Gabriel Aurioles, Professor
Construction Management Department
Florida International University
University Park VH230
107th Avenue and 8th Street
Miami, FL 33199
305-348-3542
   BS/MS, Construction Management

Miami-Dade Community College, Wilfred J. Muniz, Director
Fire Science Technology
Academy of Science
Miami-Dade Community College
11380 NW 27th Avenue
Miami, FL 33167
305-237-1400
   AS, Fire Science Technology
   AS, Fire Science Administration

Hillsborough Community College, Keith Day, Coordinator
Fire Safety Department
Hillsborough Community College
RD Box 5096
Tampa, FL 33675-5096
813-253-7628
    AS, Fire Science Technology

University of Florida, Richard Coble, PhD, Associate Professor
M.E. Rinker Senior
School of Building Construction
University of Florida
FAC 100/BON
Gainesville, FL 32611-2032
352-392-7521
    MS, Building Construction, Concentration in Construction Safety

University of Florida, Dr. Joseph J. Delsino, Chair
Department of Environmental Engineering Sciences
University of Florida
PO Box 116450
Gainesville, FL 32611-6450
352-392-0841
    BS/MS/PhD, Environmental Engineering

University of Georgia, Harold Barnhart, Coordinator
Environmental Health Science
University of Georgia
Room 206, Dairy Science Building
Athens, GA 30602-2102
706-542-2454
    BS, Environmental Health Science

Georgia Institute of Technology, Dr. Leland Riggs, Associate
    Director/Academic
Graduate Program of Environmental Engineering
School of Civil Engineering
Georgia Institute of Technology
790 Atlantic Drive
Atlanta, GA 30332
404-894-2000
    MS/PhD, Environmental Engineering

University of Hawaii, Arthor Kodama, Department Chair
Environmental and Occupational Health Program
Department of Public Health Sciences

Appendix E: Assistance through local colleges and universities 219

School of Public Health
University of Hawaii
1960 East-West Road
Honolulu, HI 96822
808-956-7425
   MS/MPH, Environmental Health

Southern Illinois University — Carbondale, Keith Contor,
   Associate Professor
Department of Technology
Southern Illinois University
Carbondale, IL 62901
618-536-3396
   BS, Industrial Technology

University of Illinois — Chicago, Dr. William Hallenbeck, Director
Industrial Hygiene Programs Environmental and Occupational
   Health Sciences
School of Public Health West
University of Illinois — Chicago
2121 W Taylor
Chicago, IL 60612
312-996-8855
   MS/PhD, Safety Engineering
   MS/PhD, Environmental Health
   MS/PhD, Industrial Hygiene
   MS/PhD, Industrial Safety

Northern Illinois University, Earl Hansen, Chair
Department of Technology
Northern Illinois University
Still Hall, Room 203
DeKalb, IL 60115-1349
815-753-0579
   BS, Industrial Technology, Concentration in Safety
   MS, Industrial Management, Concentration in Safety or Industrial
      Hygiene
   PhD, Education, Concentration in Safety

Western Illinois University, Dan Sigwart, Professor
Health Sciences Department
Western Illinois University
402 Stipes Hall
Macomb, IL 61455
309-298-2240
   BS, Health Science Minor in Industrial Safety

Illinois State University, Edmond Corner, Director
Safety Studies
Department of Health Sciences
College of Applied Science and Technology
Illinois State University, Mail Code 5220
Normal, IL 61790-5220
309-438-8329
    BS, Safety
    BS, Environmental Health

University of Illinois — Urbana-Champaign, Vernon Snoeyink, Supervisor
Environmental Engineering and Science Program
Civil Engineering Department
3230 Newmark CE Lab
University of Illinois
205 N Matthews
Urbana, IL 61801
217-333-6968
    BS/MS, Civil Engineering, Environmental Emphasis

Indiana University, James W. Crowe, Chair
Hazard Control Program
Applied Health Science/HCP
School of Health, PE and Recreation
Indiana University — Bloomington
HPER 116
Bloomington, IN 47405
812-855-2429
    AS, Hazard Control
    BS, Occupational Safety and Health
    MS, Safety Management
    HSD, Safety Education

Indiana State University, John Doty, Chair
Industrial Health and Safety Management Program
Applied Health Science Department
School of Health, PE and Recreation
Indiana State University
Terre Haute, IN 47809
812-237-3079
    BS, Safety Management
    BS, Environmental Health
    MS, Health and Safety

Purdue University, Dr. Paul Ziemer, Department Head
School of Health Sciences
Purdue University
1163 Civil Engineering Building
West Lafayette, IN 47907
317-494-1392
   BS, Environmental Health
   BS, Environmental Engineering
   BS/MS/PhD, Industrial Hygiene
   BS/MS/PhD, Health Physics

Purdue University, William E. Field, Professor
Department of Agricultural Engineering
Purdue University
1146 Agricultural Engineering Building
West Lafayette, IN 47907-1146
317-494-1173
   MS/PhD, Agricultural Safety and Health

Iowa State University, Jack Beno, Coordinator
Occupational Safety Program
School of Education
Iowa State University
Industrial Education Building 2
Room 122
Ames, IA 50010
515-294-5945
   BS, Occupational Safety and Health

Western Kentucky University, Donald Carter, Coordinator
Occupational Health and Safety Program
Department of Public Health
Western Kentucky University
1 Big Red Way
Bowling Green, KY 42101
502-745-5854
   AS, Occupational Safety and Health
   BS, Industrial Technology, Concentration in Occupational Safety and Health

Morehead State University, Dr. Brian Reeder, Coordinator
Department Biological-Environmental Sciences
Morehead State University
Morehead, KY 40351
606-783-2945
   BS, Environmental Studies

Murray State University, David G. Kraemer, Chair
Occupational Safety and Health Department
Murray State University
PO Box 9
Murray, KY 42071
502-762-2488
   BS/MS, Occupational Safety and Health

Eastern Kentucky University, Larry Collins, Coordinator
Fire and Safety Engineering Technology Program
Loss Prevention and Safety Department
College of Law Enforcement
Eastern Kentucky University
220 Stratton Building
Richmond, KY 40475
606-622-1051
   AA, Fire and Safety
   BS, Fire and Arson
   BS, Industrial Risk Management
   BS, Fire Protection Administration
   BS, Fire Protection Engineering Technology
   BS, Insurance and Risk Management
   MS, Loss Prevention and Safety

Louisiana State University, Lalit Verma, Department Head
Department of Agriculture-Engineering
Louisiana State University
Baton Rouge, LA 70803
504-388-3153
   BS, Industrial and Agricultural Technology

Nicholls State University, Michael Flowers, Program Coordinator
Petroleum Services Department
Nicholls State University
PO Box 2148
University Station
Thibodaux, LA 70301
504-448-4740
   AS, Petroleum Safety

University of Southwestern Louisiana, Thomas E. Landry,
   Associate Professor
Department of Industrial Technology
University of Southwestern Louisiana
PO Box 42972
Lafayette, LA 70504

318-482-6968
   BS, Industrial Technology, Concentration in Safety

Central Maine Technical College, Patricia Vampatella, Assistant Dean
Occupational Health and Safety Department
Central Maine Technical College
1250 Turner Street
Auburn, ME 04210
207-784-2385
   AAS, Applied Science

Johns Hopkins University, Dr. Patrick Breysse, Director
School of Hygiene and Public Health Environmental Sciences
Johns Hopkins University
615 N Wolfe Street
Baltimore, MD 21205
410-955-3602
   MHS/PhD, Environmental Health, Engineering and Safety Sciences
   MHS/PhD, Industrial Hygiene and Safety Sciences

Salisbury State University, Elichia A. Venso, PhD, Assistant Professor
Environmental Health Department
Salisbury State University
Salisbury, MD 21801
410-543-6490
   BS, Environmental Health

University of Maryland, Dr. Steven Spivak, Chair
Department of Fire Protection Engineering Room 0151
Engineering Classroom Building A
James Clark School of Engineering
Glenn L. Martin Institute of Technology
University of Maryland
College Park, MD 20742-3031
301-405-6651
   BS, Fire Protection Engineering
   MS, Fire Protection Engineering
   ME, Fire Protection Engineering

North Shore Community College, Frank Ryan, Chair
Fire Protection Safety Department
North Shore Community College
1 Ferncroft Road
Danvers, MA 01923
508-762-4000, ext. 5562
   AA, Fire Protection Safety Technology

University of Massachusetts, Dr. Michael Ellenbecker, Coordinator
Work Environments Department
University of Massachusetts
1 University Avenue
Lowell, MA 01854
508-934-3250
    MS, Engineering, Concentration in Industrial Hygiene and Ergonomics
    MS/ScD, Engineering, Concentration in Work Environments and
        Safety Ergonomics

Tufts University, John Kreilfeldt, Professor
Human Factors Program
Mechanical Engineering Department
College of Engineering
Tuffs University
Anderson Hall
Medford, MA
617-628-5000, ext. 2209
    BS, Engineering Psychology
    MS/PhD, Human Factors

Worcester Polytechnic Institute, David Lucht, Director
Fire Protection Engineering
Center for Fire Safety Studies
Worcester Polytechnic Institute
100 Institute Road
Worcester, MA 01609
508-831-5593
    MS/PhD, Fire Protection Engineering

Henry Ford Community College, Sally Goodwin, Director
Management Development Division
Henry Ford Community College
22586 Ann Arbor Trail
Dearborn Heights, MI 48127
313-730-5960
    AS, Fire Science
    AA, Property Assessment

Wayne State University, Dr. David Bassett, Chair
Occupational and Environmental Health Sciences
College of Pharmacy and Allied Health
Wayne State University
628 Shapero Hall
Detroit, MI 48202
313-577-1551
    MS, Occupational and Environmental Health

Madonna University, Florence Schaldenbrand, Chair
Physical and Applied Sciences
College of Science and Mathematics
Madonna University
36600 Schoolcraft Road
Livonia, MI 48150-1173
313-591-5110
   AS/BS, Occupational Safety, Health and Fire Science

Central Michigan University, Louis Ecker, Professor
Department of Industrial and Engineering Technology
Central Michigan University
Mount Pleasant, MI 48859
517-774-6443
   BS, Applied Science
   Minor in Industrial Safety
   MS, Industrial Management and Technology

Oakland University, Dr. Sherryl Schutz, Director
Industrial Health Program
School of Health Sciences
Oakland University
Rochester, MI 48309-4401
313-370-4038
   BS, Industrial Safety

Grand Valley State University, Dr. Eric Van Fleet, Director
Occupational Safety and Health Program
School of Health Sciences
Grand Valley State University
1 Campus Drive
Allendale, MI 49401-9403
616-895-3318
   BS, Occupational Safety and Health

University of Michigan — Ann Arbor, Frances Bourdas,
   Graduate Program Assistant
Industrial Operations Engineering
University of Michigan — Ann Arbor
1205 Beal Avenue, IDE Building
Ann Arbor, MI 48109-2117
313-764-6480
   BS/MS/MSE/PhD, Industrial and Operations Engineering
   MS, Engineering/Occupational Ergonomics

University of Michigan — Ann Arbor, Dr. Richard Garrison, Director
Environmental and Industrial Health Department
School of Public Health
University of Michigan — Ann Arbor
Ann Arbor, Ml 48109
313-764-2594
  MS/MPH, Industrial Hygiene
  MPH/PhD, Environmental Health

Ferris State University, Lori A. Seller, Assistant Professor
College of Applied Health Sciences
Ferris State University
200 Ferris Drive
Big Rapids, Ml 49307
616-592-2307
  BS, Industrial Safety and Environmental Health

University of Minnesota — Duluth, Bernard DeRobels, Director
Master of Industrial Safety Program
Department of Industrial Engineering
University of Minnesota — Duluth
105 Voss-Kovach Hall
Duluth, MN 55812
218-726-8117
  MIS, Industrial Hygiene
  MIS, Industrial Safety

University of Minnesota, Kathy Soupir, Coordinator
Environmental and Occupational Health
University of Minnesota
School of Public Health
RD Box 807, UMHC
Minneapolis, MN 55455
612-625-0622
  MS/PhD, Environmental Health

University of Southern Mississippi, Dr. Emmanuel Ahua,
  Program Director
Center for Community Health
College of Health and Human Sciences
University of Southern Mississippi
Box 5122 Southern Station
Hattiesburg, MS 39406-5122
601-266-5437
  MPH, Public Health, Concentration in Occupational and
    Environmental Health

Appendix E: Assistance through local colleges and universities

Central Missouri State, Dr. John J. Prince, Department Head
Safety Science and Technology Department
Central Missouri State
Humpreys Building, Room 325
Warrensburg, MO 64093
816-543-4626
  BS, Safety Management
  BS/MS, Industrial Hygiene
  MS, Transportation Safety
  MS, Fire Science
  MS, Public Service Administration
  MS, Security
  MS, Industrial Safety Management
  ED, Safety

St. Louis Community College — Forest Park, Emil Hrhacek, Coordinator
Municipal Services
St. Louis Community College — Forest Park
5600 Oakland
St. Louis, MO 63110
314-644-9310
  AA, Fire Protection Safety

Montana Tech, Julie B. Norman, CIH, Department Head
Occupational Safety and Health/Industrial Hygiene Department
Montana Tech
1300 W Park Street
Butte, MT 59701
406-496-4393
  AS, Occupational Safety and Health
  BS, Occupational Safety and Health
  BS, Environmental Engineering
  MS, Industrial Hygiene

University of Nebraska — Kearney, Darrel Jensen, Director
Nebraska Safety Center
University of Nebraska — Kearney
West Center
Kearney, NE 68849
308-234-8256
  BS, Safety Education
  BS, Occupational Safety and Health
  BS, Transportation Safety
  BS, Driver Education

Community College of Southern Nevada, Sonny Lyerly, Chair
Department Mathematics, Health and Human Services
Community College of Southern Nevada
6375 W Charleston
Las Vegas, NV 89102
702-643-6060, ext. 439
   AAS, Fire Science Technology

Keene State College, David Buck, Director
Safety Center
Keene State College
229 Main Street
Keene, NH 03431
603-358-2977
   AS, Chemical Dependency
   BS, Industrial Safety
   BS, Occupational Safety and Health

Camden County College, Matthew Davies, Coordinator
Information Services
Camden County College
RD Box 200
Blackwood, NJ 08012
609-227-7200, ext. 251
   AA, Occupational Safety
   AA, Fire Science

Rutgers, The State University of New Jersey, Frank Haughey, Director
Radiation Science Program Rutgers, The State University of New Jersey
Building 4087, Livingston Campus
New Brunswick, NJ 08093
908-932-2551
   BS/MS, Radiation Science

New Jersey Institute of Technology, Howard Gage,
   Director/Associate Professor
Occupational Safety and Health Department of Mechanical and
   Industrial Engineering
New Jersey Institute of Technology
University Heights
Newark, NJ 07102
201-596-3653
   MS, Occupational Safety and Health

Thomas Edison State College, Janice Touver, Admissions
Applied Science and Technology

Appendix E: Assistance through local colleges and universities

Thomas Edison State College
101 W State Street
Trenton, NJ 08608-1176
609-984-1150
   AS/BS, Fire Protection Science
   AS/BS, Environmental Science and Technology
   AS/BS, Industrial Engineering Technology

New Mexico Institute of Mining & Technology, Dr. Clint Richardson,
   Associate Professor
Department of Mineral and Environmental Engineering
New Mexico Institute of Mining and Technology
Campus Station
801 Leroy
Socorro, NM 87801
505-835-5345
   BS, Environmental Engineering

Broome Community College, Francis Short, Chair
Special Career Programs Department
Broome Community College
RD Box 1017
Binghamton, NY 13902
607-778-5000
   AAS, Fire Protection Safety

Mercy College, Dr. Joe Sullivan, Chair
Criminal Justice and Public Safety Department
Mercy College
Social Science Building
555 Broadway
Dobbs Ferry, NY 10522
914-674-7320
   BS, Public Safety Certificates: Fire Science, OSHA, Public Safety,
      Private Security

New York University, Katie B. Shadow, Graduate Coordinator
Environmental Health Sciences Program
Nelson Institute of Environmental Medicine
New York University
A.J. Lanza Laboratories
Long Meadow Road
Tuxedo, NY 10987
914-351-5480
   MS, Occupational and Industrial Hygiene
   PhD, Environmental Health Sciences

State University of New York, College of Technology, John Tiedemann
   Department Head
Department of Industrial Technology-Facility Management Technology
State University of New York, College of Technology
Route 110
Farmingdale, NY 11735
516-420-2326
   BS, Industrial Technology

Columbia University, Anne Hutzelmann, Administrative Assistant
Division of Environmental Sciences
Columbia University
188th Street
New York, NY 10032
212-305-3464
   MS/DrPH, Public Health

University of Rochester, Mary Wahlman, Coordinator
Department of Biophysics
School of Medicine
University of Rochester
Rochester, NY 14642
716-275-3891
   MS, Environmental Studies
   MS, Industrial Hygiene

University of North Carolina — Chapel Hill, David Leith,
   Program Director
Environmental Sciences and Engineering
School of Public Health
University of North Carolina — Chapel Hill
201 Columbia Street
Chapel Hill, NC 27599-7400
919-966-3844
   BSPH/PhD, Environmental Science and Policy
   MS, Public Health

University of North Carolina — Chapel Hill, Larry Hyde,
   Deputy Director
Research Center
University of North Carolina — Chapel Hill
109 Conners Drive #1101
Chapel Hill, NC 27514
919-962-2101
   Continuing Education in Occupational Health and Safety
   Private Seminars

Appendix E: Assistance through local colleges and universities 231

Central Piedmont Community College, Andy Nichols, Director
Industrial Safety
Central Piedmont Community College
RD Box 35009
Charlotte, NC 28235-5009
704-342-6582
   AA, Industrial Safety

Western Carolina University, Robert Dailey, Coordinator
Occupational Safety Program
Industrial and Engineering Technology Department
Western Carolina University
226 Belk Building
Cullowhee, NC 28723
704-227-7272
   BS, Electronics Engineering Technology
   BS, Industrial Technology
   BS, Manufacturing Engineering Technology
   BS, Industrial Distribution
   MS, Technology

North Carolina A&T State University, Dillip Shah, Coordinator
Department of Construction Management and Safety
North Carolina A&T State University
Price Hall, Room 124
Greensboro, NC 27411
919-334-7586
   BS, Occupational Safety and Health

East Carolina University, Dr. Mark Friend, Program Director
Department of Industrial Technology
East Carolina University
105 Flanagan
Greenville, NC 27858
919-328-4249
   BS, Environmental Health, Option in Industrial Hygiene
   MSIT, Occupational Safety

North Carolina State University, Richard G. Pearson, Professor
Department of Industrial Engineering
North Carolina State University
Box 7906
Raleigh, NC 27695
919-515-6410
   PhD, Industrial Engineering, Concentration in Ergonomics

North Dakota State College of Science, Linda Johnson, Instructor
North Dakota State College of Science
800 N 8th Street
Wahpeton, ND 58076
701-671-2202
  AS, Industrial Hygiene
  AS, Occupational Health and Safety

University of Akron, Dr. David H. Hoover, Program Head
Fire Protection Program
Division of Public Service Technology
The University of Akron
Akron, OH 44325-4304
216-972-7789
  AAS, Fire Protection Technology, 2+2 Option in Technical Education
  BS, Fire Protection

University of Cincinnati, William M. Kraemer Director,
  College of Applied Science
University of Cincinnati
2220 Victory Parkway, ML 103
Cincinnati, OH 45206
513-556-6583
  AAS, Fire Science Technology
  BS, Fire Science Engineering

University of Cincinnati, Dr. Rod Simmons,
  Assistant Research Professor
Department of Mechanical, Industrial, and Nuclear Engineering
University of Cincinnati
Mail Location 116
Cincinnati, OH 45221-0116
513-556-2738
  MS/PhD, Industrial Engineering, Concentration in Occupational Safety

Stark Technical College, Cameron H. Speck, Program Developer
Safety/Risk Management
Continuing Education
Stark Technical College
6200 Frank Avenue, N.W.
Canton, OH 44720
216-494-6170
  AS, Engineering Technology
  AS, Allied Health

Appendix E: Assistance through local colleges and universities 233

Wright State University, Allan Burton, Director
Environmental Health Sciences Program
Biological Sciences Department
Wright State University
Colonel Glenn Highway
Dayton, OH 45435
513-873-2655
  BS, Environmental Sciences

Wright State University, Dr. Jennie Gallimore, Associate Professor
Department of Biomedical and Human Factors Engineering
College of Engineering
Wright State University
207 Russ Center
Dayton, OH 45435
513-873-5044
  BS, Human Factors Engineering

East Central University, Dr. Paul Woodson, Chair
Environmental Science Program
Physical and Environmental Sciences Department
East Central University
Ada, OK 74820
405-332-8000, ext. 547
  BS, Environmental Science, Concentrations in Environmental Health,
    Industrial Hygiene and Environmental Management

University of Central Oklahoma, Dr. Lou Ebrite, Department Chair
Occupational and Technology Education Department
College of Education
University of Central Oklahoma
100 N University Drive
Edmond, OK 73034-0185
405-341-5009
  BS, Industrial Safety

University of Oklahoma, Deborah Imel Nelson, Program Head
Civil Engineering and Environmental
Science Department
University of Oklahoma
202 W Boyd Street, Room 334
Norman, OK 73109
405-325-5911
  MS, Environmental Science

University of Oklahoma — Oklahoma City, Robert Nelson,
   Associate Professor
Occupational and Environmental Health Department
University of Oklahoma — Oklahoma City
PO Box 26901
Oklahoma City, OK 73190
405-271-2070
   MS/MPH, Environmental Management
   MS/MPH, Environmental Toxicology
   MS/MPH, Industrial Hygiene

Southeastern Oklahoma State University, Robert Semonisck, PhD,
   Professor, Safety
School of Applied Science and Technology
Southeastern Oklahoma State University
Station A
Durant, OK 74702
405-924-0121, ext. 2464
   BS, Occupational Safety and Health

Oklahoma State University, Dr. Don Adams, Coordinator
Fire Protection and Safety Technology Department
303 Campus Fire Station
Oklahoma State University
Stillwater, OK 74078
405-744-5639
   BS, Fire Protection and Safety Engineering Technology

Mount Hood Community College, Dr. David Mohtasham, Coordinator
ESHM Program
Mount Hood Community College
Route 26,000 SE Stark Street
Gresham, OR 97030
503-667-7440
   AAS, Environmental Safety and Hazardous Materials Management

Southwestern Oregon Community College, Darryl Saxton, Coordinator
Fire Science Program
Southwestern Oregon Community College
Coos Bay, OR 97420
503-888-2525
   AAS, Fire Protection

Oregon State University, Dave Lawson, Associate Professor
Safety Studies Program
Department of Public Health

Appendix E: Assistance through local colleges and universities 235

College of Health and Human Performance
Oregon State University
Waldo Hall, Room 256
Corvallis, OR 97331-6406
503-737-2686
   BS, Environmental Health and Safety
   MS, Safety Management
   MS, Environmental Health Management, Concentration in
     Occupational Safety
   PhD, Health

Indiana University of Pennsylvania, Dr. Robert Soule, Chair
Safety Science Department
College of Health and Human Science
Indiana University of Pennsylvania
117 Johnson Hall
Indiana, PA 15705
412-357-3019
   BS/MS, Safety Sciences

Millersville University of Pennsylvania, Dr. Paul Specht, Coordinator
Department of Industry and Technology
Millersville University of Pennsylvania
PO Box 1002
Millersville, PA 17551
717-872-3981
   BS, Occupational Safety and Hygiene Management

Northampton Community College, Kent Zimmerman, Program
   Director
Safety, Health and Environmental Technology
Northampton Community College
3835 Green Pond Road
Bethlehem, PA 18017
610-861-5590
   AAS, Applied Science, Concentrations in Safety, Health and
     Environmental Technology

Slippery Rock University of Pennsylvania, Dr. Joseph Calli, Chair
Allied Health Department
Slippery Rock University of Pennsylvania
Behavioral Science Building, Room 208
Slippery Rock, PA 16057
412-738-2017
   BS, Safety and Environmental Management

Francis Marion University, Dr. W.H. Breazeale, Department Head
Department of Chemistry and Physics
Francis Marion University
RD Box 100547
Florence, SC 29501
803-661-1440
  BS, Health Physics

University of South Carolina — Columbia, Dr. Edward Oswald, Professor
Department of Environmental Health
School of Public Health Sciences
University of South Carolina — Columbia
Health Sciences Building, Room 311B
Columbia, SC 29208
803-777-4120
  MSPH/MPH/PhD, Occupational Health, Environmental Duality and Hazardous Materials Management

East Tennessee State University, Creg Bishop, Interim Chair
Environmental Health Department
College of Public and Allied Health
East Tennessee State University
Johnson City, TN 37614
615-929-4268
  BS/MS, Environmental Health
  Minor in Safety

Middle Tennessee State University, Dr. Richard Redditt, Professor
Industrial Studies Department
Middle Tennessee State University
PO Box 19
Murfreesboro, TN 37132
615-898-2776
  MS, Industrial Studies, Concentration in Safety

University of Tennessee — Knoxville, Charles Hamilton, Chair
Health, Leisure and Safety Department
University of Tennessee — Knoxville
1914 Andy Holt Drive
Knoxville, TN 37996-2700
615-974-6041
  BS/MS/EdD/PhD, Health Education
  MS/EdS, Safety Education
  MS, Public Health

Appendix E: Assistance through local colleges and universities 237

Lamar University, Dr. Victor Zalcom, Department Chair
Industrial Engineering
Lamar University
PO Box 10032-LUS
Beaumont, TX 77710
409-880-8804
  BS, Industrial Technology
  BS, Industrial Engineering

University of Houston — Clearlake, Dr. Dennis Casserly,
  Associate Professor
Division of Natural Sciences
University of Houston — Clearlake
2700 Bay Area Boulevard
Houston, TX 77058
713-283-3775
  BS, Environmental Science

Texas A & M University, Dr. James Rock, Associate Professor
Safety Division
Nuclear Engineering
Texas A & M University
College Station, TX 77843-3133
409-862-4409
  BS/MS, Safety Engineering
  BS/MS, Industrial Hygiene
  BS/MS, Health Physics

Texas Tech University, Dr. Mica Endsley, Assistant Professor
Department of Industrial Engineering
Texas Tech University
PO Box 43061
Lubbock, TX 79409
806-742-3543
  BS/MS/PhD, Industrial Engineering, Concentration in Ergonomics

Sam Houston State University, Dr. James R. DeShaw, Program Head
Department of Biological Sciences
Sam Houston State University
PO Box 2116
Huntsville, TX 77341-2116
409-294-1020
  BS, Environmental Sciences

San Jacinto College Central, Gary M. Vincent, Chair
Division of Health Science
Health and Safety Technology Department
San Jacinto College Central
8060 Spencer Highway
Pasadena, TX 77501-2007
713-476-1834
   AA, Occupational Health and Safety Technology

The University of Texas at Tyler, Dr. W. Clayton Allen, Chair
The University of Texas at Tyler
School of Education and Psychology
Department of Technology
3900 University Boulevard
Tyler, TX 74799
903-566-7331
   BS/MS, Technology, Concentration in Industrial Safety

Texas State Technical College, David Day, Department Chair
Occupational Safety and Health Department
Texas State Technical College
3801 Campus Drive
Waco, TX 76705
817-867-4841
   AAS, Occupational Safety and Health
   AAS, Hazardous Materials Management
   AAS, Radiation Protection Technician

University of Utah, Donald S. Bloswick, Associate Professor
Mechanical Engineering Department
University of Utah
3209 MEB
Salt Lake City, UT 84112
801-581-4163
bloswick@me.mech.utah.edu
   MS/ME/PhD, Mechanical Engineering, Concentration in Ergonomics
      and Safety MPH/MSPH, Public Health, Concentration in
      Ergonomics and Safety

Virginia Commonwealth University, Michael McDonald, Coordinator
Safety and Risk Administration Program
Justice/Risk Administration Department
School of Community and Public Affairs
Virginia Commonwealth University
913 W Franklin Street
Richmond, VA 23284

804-828-6237
  BS, Safety and Risk Control Administration

Virginia Tech University, Tom Dingus, Professor
Department of Industrial Engineering
Virginia Tech University
302 Whittemore Hall
Blacksburg, VA 24061
540-231-8831
  MS, Safety Engineering

Central Washington University, Ronald Hales, Professor
Industrial Engineering
Technology Department
Hebeler Hall
Central Washington University
Ellensburg, WA 98926
509-963-3218
  BS, Loss Control Management
  Minor in Traffic Safety
  Minor in Loss Control Management

University of Washington, Mary Lou Wager, Graduate Program Assistant
Environmental Health Department School of Public Health and
  Community Medicine
Mail Stop SC-34
University of Washington
Seattle, WA 98195
206-543-3199
  MS/PhD, Industrial Hygiene and Safety

Fairmont State College, John Parks, Safety Coordinator
Technology Division
Fairmont State College
Locust Avenue
Fairmont, WV 26554
304-367-4633
  BS, Safety Engineering Technology

West Virginia University, Terrence Stobbe, Director
Department of Industrial Engineering
College of Engineering
West Virginia University
RD Box 6101
Morgantown, WV 26506-6101
304-293-4607
  MS, Occupational Hygiene and Occupational Safety

West Virginia University, Daniel E. Della-Guistina, Chair
Department of Safety and Environmental Management
West Virginia University
PO Box 6070, COMER
Morgantown, WV 26506
304-293-2742
   MS, Safety and Environmental Management

Marshall University, Keith Barenklau, Program Director
Safety Technology Department
Gullickson Hall, Room 3
College of Education
Marshall University
Huntington, WV 25755-2460
304-696-4664
   BS/MS, Safety Technology, Occupational Safety Option
   MS, Safety Technology, Safety Management Option
   MS, Mine Safety

University of Wisconsin — Eau Claire, Dale Taylor, Chair
Department of Allied Health Professions
University of Wisconsin — Eau Claire
Eau Claire, WI 54702-4004
715-836-2628
   BS, Environmental and Public Health
   MS, Environmental and Public Health

University of Wisconsin — Stout, John Olson, Director
Safety and Loss Control Center
Industrial Management Department
University of Wisconsin — Stout
205 Communications Center
Menomonie, WI 54751
714-232-2604
   MS, Occupational Safety and Health

University of Wisconsin — Platteville, Roger Hauser, Professor
Industrial Studies Department
University of Wisconsin — Platteville
309 Pioneer Tower
Platteville, WI 53818
608-342-1187
   BS/MS, Industrial Technology, Management and Occupational Safety,
     Concentration in Safety

University of Wisconsin — Steven's Point, Dr. Ann Abbott, Director
School of HPERA
University of Wisconsin — Steven's Point
131 Quandt
Steven's Point, WI 54481
715-346-4420
  BS, Health Promotion and Safety Health Protection
  Minor in Safety

University of Wisconsin — Whitewater, Jerome W. Witherill, Chair
Department Safety Studies
University of Wisconsin — Whitewater
800 W Main Street
Whitewater, WI 53190
414-472-1117
  BS/MS, Safety Majors in Institutional Safety, Occupational Safety and
    Traffic Safety[1]

---

[1] American Society of Safety Engineers, 1996-97 Survey of College and University Safety and Related Degree Programs.

Index

A

Action Plan, 53
Agriculture, 10
Aircraft
    Crashes/incidents, 16–17
Americans Disabilities Act (ADA), 45, 121
    Title I (Employment Provisions), 143–146
    Title II (Public Services), 145–147
    Title III (Public Accommodations), 147–148
    Title IV (Telecommunications), 148–149
    Title V (Miscellaneous Provisions), 149–150
American National Standards Institute (ANSI), 98
Analysis, 3, 4
Animal Care/Control Agency, 207–209
Ardentes, 8
Assessment, 1, 3, 15
    Checklist, 50–51
    Concerns list, 30
    Monetary, 4
    Subjective, 2
    Workable, 4
Assets, 1
    Local, 39
    Safeguard, 12
Atmospheric, 13
Avalanches, 14
Aviation centers, 17

B

Basaltic composition, 8
Basic Assessment List, 3
Bio-terrorism, 21, 26
Buffalo Creek, 18

C

Caldera, 8
Catastrophic, 4
CD ROM, 76
Chemist
    John Walker, 10
Chief Executive Officer, 190
Circum-Pacific belt, 6
Collapse, 8
Combustible, 9
Communication
    Coordinator, 193–194
    Effective, 61
    Functional positions, 62
    System of management, 61, 93
Compliance, 33
Compression, 10
Comptroller's Office, 209
Construction equipment, 92
Contingency plan, 47
Convergent plate, 6
Corporate veil, 157
County Coroner/Medical Examiner, 197
Counterclockwise, 13
Critical stress debriefing, 92
Crustal deformation, 7
Cultivation
    Slash and burn, 10
Cumulonimbus cloud, 13
Customize, 1
Cyber-terrorism, 21, 25

D

Dams
    Earthen, 18
Damages
    Compensatory, 132
    Long term, 16
    Potential, 3
    Punitive, 132
Day to Day Operations, 4
Death, 2, 10
Debris, 92

Decision-making, 3
De Minimis violations, 102–103
Detrimental, 3
Discrimination
    Intentional, 133
Duration, 12
Dilatancy, 7
Direct link society, 21
Disability
    ADA, 121
    Issues, 121–123
Disaster
    Criminal Sanctions, 155
    Factors, 151–153
    Personal, 155
    Pre-planning, 43
    Preparedness assessment components, 151
    Preparedness plan, 1, 2, 19
    Sources/Assistance List, 213–241
    Web sites, 183–188
Disease
    Environmental, 47
    Tuberculosis, 14
    Outbreaks, 14
Distribution, 6
Disturbance, 5
Dividends, 4
Dollar losses, 2
Down-sizing, 23, 87

E

Earth
    Crust, 6
Earthquakes
    Definition, 5
    Intraplate, 6
Education Department, 206–207
Efficacy, 3
Elastic
    Waves, 5
    Strain, 6
Emergency Planning and Community Right-to-Know Act (EPCRA), 30
    Medical Services/Agencies, 198–199
    Plan, 1–2, 42
    Operations Centers, 63–64
    Response planning team, 67–70
Emergency Program Manager, 191–193
Emerging risks, 21
Environment, 1, 7, 15
Environmental Protection Agency (EPA), 29
    Website, 29
Eons, 10

Epicenter, 7
Epicentral distance, 6
Epidemic
    Major, U.S., 14
Estimate, 3
Equal Employment Opportunity Commission (EEOC), 123
Equilibrium, 6
Event, 3
Evacuation, 12
    Alarm, 45
    Stations, 45
Expending, 3
External actions, 46

F

Fact of life, 16
Fault zone, 7
Federal Communication Commission, 130
Federal Emergency Management Agency (FEMA), 29
Feedback mechanisms, 81
Fertilizer, 10
Fire
    Chemistry, 37
    Definition, 9
    Drills, 10
    Explosions (recent), 11–12
    Friction method, 10
    Fire Ground Command System (FGCS), 61
    History, 10
    Piston, 10
    Plow, 10
    Protection activities, 11
    Saw, 10
    Techniques, 10
First Responder Awareness Level, 41
Flint, 10
Floods, 14
    Disasters, 15
Formation, 8
Friction match, 10
Fruition, 19

G

General Duty Clause, 100
General Services Department, 209–212
Geology, 7
Global economy, 16
Governmental
    Reactions, 95

Index

Regulations/Primary agencies, 29

H

Hazards, 5
Hazardous Material Response Team, 42
Hazmat, 53
Historical view, 5
Hominids, 10
Horizontal axis, 6
Hurricanes
    Define, 11
    Eye, 12
    Recent, 12
    Stages, 11

I

Incident, 2
Incident Command System (ICS), 61
Incident Management System (IMS), 61
Identify, 1
Industrial
    Revolution, U.S., 23
    Safety, 158
    Society, 16
Industry, 11
Ignimbrites, 8
Inherent, 5
Injury, 2
Insurance, 3
    Companies, 92
Investment, 4
Island arcs, 6
Internal actions, 43
    Chain of command chart, 43
    Communications, 64–65
    Employee emergency contact, 43

K

Kenya, 10
Kilavea, 8
Kilometers, 8

L

Landform, 8
Landslide, 7
Lava, 8
Legal
    Department, 207
    Issues, 115–116
Learning
    Auditory, 74
    Doing, 74
    Visual, 74
Life, 2–3
Lightning, 10
Local Emergency Planning Committee (LEPC), 40
Location, 1
Logarithmic scale, 7
Long-term, 16
Loss of life, 18

M

Magnitude, 1
Magma, 8
Management team, 3, 4, 16, 18
    Command, 61
    System, 61
Market value, 46
Mass Care
    Coordinator, 194–195
    Facility Manager, 195–196
Mass casualties, 46
Measurement, 2
Media Control, 21, 46, 83–86
    Issues, 85–86
Methodology, 2
Microscopic vision, 5
Mid-ocean ridges, 6
Mines
    Explosions, 18
    Open pit, 18
    Underground, 18
Miranda Rights, 168
"Mission Statement", 37, 53
Military Department, 207
Modular system, 61
Monetary
    Liabilities, 101
    Terms, 3
Mount St. Helen's, 8
Mudslide, 7
Murphy's Law, 19

N

National Fire Protection Association, 35
National Institute of Occupational Safety and Health (NIOSH), 22, 96, 181
Natural disaster, 5
Neolithic, 10
Nuees, 8
Numerical formula, 2

O

Occupational Safety and Health
  Administration (OSHA), 95–96
  Guidelines, 24
  Inspection Checklist, 171–172
  Standards Site List, 30–31
  Website, 29
Oil
  Major spills, 16
Operations
  Day to day, 4
Origin, 6
Oscillation, 6

P

Pamphlet 13 Standard, 35
Pandora's Box, 21
Peking man, 10
Petroleum, 15
Phenomena, 6, 22
  Precursor, 7
Phosphorous sulfate, 10
Plan
  Emergency and Disaster, 1, 2, 4
  Preventative, 4
  Proactive, 4
Planet earth, 5
Plate tectonics theory, 6
Postulates, 6
Potential
  Loss, 16
  Risk, 1–3, 18
Prehistoric, 10
Premiums, 4
Pre-Planning Disaster
  Internal actions, 43
  External actions, 46
Primitive, 10
Proactive prevention program, 19
Probability, 2–3
Process, 1
Process Safety Management Standard, 40
Product tampering, 36
Productivity, 3
Profitability, 3
Propagation, 6
Property, 2, 3, 10
Public
  Health Coordinator, 196–197
  Information Officer, 46, 199–200
  Works, 205–206

Pyrites, 10
Pyroclastic debris, 7

R

Radon, 7
Railroad
  Accidents, 19
Reasonable accommodation, 126
Recording station, 6
Refracted, 6
Rehabilitation Act, 123
Re-ignite, 10
Remote, 19
Resources, 2
  Contact List, 40–41
  Manager, 200–201
  Natural, 16
Richter scale, 7
Rift system, 6
Rigid plates, 6
Risks
  Catastrophic, 4
  Eliminating, 49
  Initial, 5
  Man, 15
  Minimizing, 49
  Natural, 5, 15
  Potential, 1, 3, 16
  Shifting, 49
  Substantial, 2, 14–15
Rock
  Layers, 6
  Masses, 6
  Molten, 7, 8

S

Safeguards, 2, 10
Safety Professionals, 1, 3
Second wave, 95
Secretary of Labor, 98–99
Seismicity
  Major zones, 6
  Occurrence, 6
Seismic waves, 6
  Amplitudes, 7
Seismologist, 7
Seismogram pattern, 6
Seismograph, 6
Shareholder factor, 87–89
Shensi Province, 7

Index

Shipwrecks
  Recent, 18
Site map, 44
Solidify, 8
"Spin", 84
Sprinkler systems
  Antifreeze, 35
  Automatic, 35
  Fire control valves, 35
  Wet pipe, 35
State Emergency Response Commission (SERC), 40
Structural Preparedness, 33
Structure, 1–2
  Man-made, 7
  Security, 35
Subjective, 2
Substantial, 2
Suppression system, 34
Synopsis, 5

T

Tectonic, 5
Terrorism
  Define, 21
  Incidents, 21–22
Thermal imaging, 34
Tidal waves, 14
Topography, 8
Tornadoes
  Define, 13
  Funnel cloud, 13
  Recent, 13–14
  Tornado belt, 13
  Twister, 13
  Vortex, 13
  Whirlwind, 13
Training
  Classroom (Formal), 73
  Computer assisted, 73
  Documentation, 79–80
  Elements, 79
  Hands on, 73, 82
  Interactive computer, 73
  Success, 71–73
Trans-Asiatic belt, 6
Tropical
  Cyclone, 11
  Depression, 11
  Storm, 11
Tsunamis, 7
Tuffs, 8
Typical Responsibilities, 189–212

U

Unlawful discrimination, 132
U.S. Bureau of Labor Statistics, 22
U.S. Postal Service
  Incidents, 23

V

Vesuvius, 8
Viability, 1
Victim, 13
Violation
  Criminal liability/penalties, 108–114
  De Minimis, 102–103
  Non-serious, 103–104
  Posting, 108
  Repeat/failure to abate, 107–108
  Serious, 104–106
  Violation/Penalty Schedule, 100
  Willful, 106–107
Viscosity, 8
Vision
  Field, 5
  Microscopic, 5
Volcanoes
  Central, 7, 8
  Cone, 8
  Definition, 7
  Eruption, 5, 8–9
  Fissure, 7

W

Warning system, 13
Waves
  Elastic, 5
  Primary, 6
  Secondary, 6
  Seismic, 6
Web Sites
  Disaster Preparedness, 183–188
Workers' Compensation System
  Commission/Board, 118
  Features, 116–118
  Plan, 120
Workplace
  Hazards, 174–175
  Rights (employee), 173–181
Workplace Violence
  Define, 23
  Terrorism, 21
  Violence, 21–22, 36
World Trade Center, 23
Written Plan, 57–58